FORAGE CROPS OF THE WORLD

Volume I: Major Forage Crops

FORAGE CROPS OF THE WORLD

Volume I: Major Forage Crops

Edited by

Md. Hedayetullah, PhD

Parveen Zaman, PhD

AAP | APPLE ACADEMIC PRESS

Apple Academic Press Inc.
3333 Mistwell Crescent
Oakville, ON L6L 0A2
Canada

Apple Academic Press Inc.
9 Spinnaker Way
Waretown, NJ 08758
USA

International Standard Book Number-13: 978-1-77188-684-0 (Hardcover)
International Standard Book Number-13: 978-1-351-16736-9 (eBook)
International Standard Book Number-13: 978-1-77188-686-4 (2-volume set, Hardcover)
International Standard Book Number-13: 978-1-351-16724-6 (2-volume set, eBook)

Library and Archives Canada Cataloguing in Publication

Forage crops of the world / edited by Md. Hedayetullah, PhD, Parveen Zaman, PhD.

Includes bibliographical references and indexes.
Content: Volume I. Major forage crops -- Volume II. Minor forage crops.
Issued in print and electronic formats.
ISBN 978-1-77188-684-0 (v. 1 : hardcover).--ISBN 978-1-77188-685-7
(v. 2 : hardcover).--ISBN 978-1-77188-686-4 (set : hardcover).--
ISBN 978-1-351-16736-9 (v. 1 : PDF).--ISBN 978-1-351-16728-4
(v. 2 : PDF).--ISBN 978-1-351-16724-6 (set : PDF)

1. Forage plants. I. Hedayetullah, Md., 1982-, editor II. Zaman, Parveen, 1989-, editor

SB193.F67 2018 633.2 C2018-903719-9 C2018-903720-2

CIP data on file with US Library of Congress

Apple Academic Press also publishes its books in a variety of electronic formats. Some content that appears in print may not be available in electronic format. For information about Apple Academic Press products, visit our website at **www.appleacademicpress.com** and the CRC Press website at **www.crcpress.com**

ABOUT THE EDITORS

Md. Hedayetullah, PhD

Md. Hedayetullah, PhD, is an Assistant Professor/Scientist and Officer In-Charge, AICRP (*All India Coordinated Research Projects*) on Chickpea, Directorate of Research, Bidhan Chandra Krishi Viswavidyalaya, Kalyani, Nadia, West Bengal. He is a former Agronomist with the NABARD, Balasore, Odisha, India. He was also formerly a Professor at the M.S. Swaminathan Institute of Agriculture Science, Centurion University of Technology and Management, Gajapati, Odisha, India, and an Assistant Professor at the College of Agriculture, Tripura, Government of Tripura, India. Dr. Hedayetullah is the author and co-author of 20 research papers, five review papers two book chapters, and one book.

Dr. Hedayetullah acquired his BS degree (Agriculture) from H.N.B. Garhwal University, Uttarakhand, India. He received his MS degree (Agronomy) from Palli Siksha Bhavana, Institute of Agriculture, Visva Bharati University, Sriniketan, West Bengal, India. He received his PhD (Agronomy) from Bidhan Chandra Krishi Viswavidyalaya, Mohanpur, Nadia, West Bengal, India. He was awarded the Maulana Azad National Fellowship Award from the University Grant Commission, New Delhi, India. He has received several fellowship grants from various funding agencies to carry out his research works during his academic career.

Parveen Zaman, PhD

Parveen Zaman, PhD, is an Assistant Director of Agriculture (Farm) at the Pulse & Oilseed Research Sub-station, Beldanga, Department of Agriculture, Government of West Bengal, India. She is author and co-author of four research papers, three review papers, and four book chapters. She acquired her BS degree (Agriculture), receiving a Gold Medal, from Bidhan Chandra Krishi Viswavidyalaya, Mohanpur, Nadia, West Bengal, India. She received MS degree (Agronomy), also with Gold Medal, from Bidhan Chandra Krishi Viswavidyalaya, Mohanpur,

Nadia, West Bengal, India. She was awarded the Maulana Azad National Fellowship Award from the University Grants Commission, New Delhi, India, and has received several fellowship grants from various funding agencies to carry out her research works during her academic career.

CONTENTS

LIST OF CONTRIBUTORS

Pintoo Bandopadhyay
Department of Agronomy, Bidhan Chandra Krishi Viswavidyalaya, Mohanpur, Nadia, India

J. Banerjee
Department of Genetics and Plant Breeding, Bidhan Chandra Krishi Viswavidyalaya, Mohanpur 741252, Nadia, West Bengal, India

Arun Kumar Barik
Department of Agronomy, Palli Siksha Bhavana, Institute of Agriculture, Visva Bharati University, Sriniketan 731236, West Bengal, India

Sonali Biswas
Department of Agronomy, BCKV, Mohanpur 741252, Nadia, West Bengal, India

W. Tampakleima Chanu
Department of Plant Pathology, College of Agriculture, CAU, Imphal, India

H. Das
GKMS, Malkangiri, Orissa University of Agriculture and Technology, Kalimela, Malkangiri, Odisha, India

M. R. Debnath
Horticulture Research Centre, Nagicherra, Tripura, India

H. Chandrajini Devi
Department of Plant Pathology, College of Agriculture, CAU, Imphal, India

R. Gajghate
Crop Production Division, Indian Council of Agriculture Research-Indian Grass and Fodder Research Institute, Jhansi 284003, Uttar Pradesh, India

Sanchita Mondal Ghosh
Department of Agronomy, Bidhan Chandra Krishi Viswavidyalaya, Mohanpur, Nadia 741252, West Bengal, India

Utpal Giri
Department of Agronomy, College of Agriculture, Lembucherra 799210, Tripura, India

G. Guleria
Department of Agronomy, Chaudhary Sarwan Kumar Himachal Pradesh Krishi Vishvavidyalaya, Palampur 176062, Himachal Pradesh, India

Jancy Gupta
Dairy Extension Division, NDRI, Karnal 132001, Haryana, India

Md. Hedayetullah
Department of Agronomy, Faculty of Agriculture, Bidhan Chandra Krishi Viswavidyalaya, Mohanpur 741252, West Bengal, India

J. K. Jadav
ICAR-CIAH, KVK-Panchmahal, Vejalpur, Godhra 389340, India

K. Jana
Department of Agronomy, Faculty of Agriculture, Bidhan Chandra Krishi Viswavidyalaya, Mohanpur 741252, Nadia, West Bengal, India

Jyotirmay Karforma
Regional Research Station (Old Alluvial zone), UBKV, Majhian, West Bengal, India

B. S. Khadda
ICAR-CIAH, KVK-Panchmahal, Vejalpur, Godhra 389340, India

Shakti Khajuria
ICAR-CIAH, KVK-Panchmahal, Vejalpur, Godhra 389340, India

R. J. Koireng
Directorate of Research, Central Agricultural University, Imphal 795004, Manipur, India

Anjani Kumar
Crop Production Division, ICAR-National Rice Research Institute, Cuttack, Odisha, India

Raj Kumar
ICAR-CIAH, KVK-Panchmahal, Vejalpur, Godhra 389340, India

U. Kumar
Crop Production Division, ICAR-National Rice Research Institute, Cuttack, Odisha, India

Champak Kumar Kundu
Department of Agronomy, Faculty of Agriculture, Bidhan Chandra Krishi Viswavidyalaya, Mohanpur 741252, Nadia, West Bengal, India

B. S. Meena
Division of Dairy Extension, Faculty of Agriculture, Karnal 132001, Haryana, India

Anupam Mukherjee
Sasya Shyamala Krishi Vigyan Kendra, Ramakrishna Mission Vivekananda University, Arapanch, Sonarpur, Kolkata 700150, India

Kanu Murmu
Department of Agronomy, Bidhan Chandra Krishi Viswavidyalaya, Mohanpur 741235, West Bengal, India

N. Nair
Department of Entomology, College of Agriculture, Tripura, Lembucherra 799210, West Tripura, India

Partha Sarathi Patra
Department of Agronomy, Uttar Banga Krishi Viswavidyalaya, Pundibari 736165, Cooch Behar, West Bengal, India

Pampi Paul
Division of Dairy Extension, National Dairy Research Institute, Karnal 132001, Haryana, India

Tarun Paul
Department of Agronomy, Uttar Banga Krishi Viswavidyalaya, Pundibari, Cooch Behar 736165, West Bengal, India

R. Poddar
Department of Agronomy, Faculty of Agriculture, Bidhan Chandra Krishi Viswavidyalaya, Mohanpur 741252, Nadia, West Bengal, India

A. M. Puste
Department of Agronomy, Bidhan Chandra Krishi Viswavidyalaya, Mohanpur 741235, West Bengal, India

A. K. Rai
ICAR-CIAH, KVK-Panchmahal, Vejalpur, Godhra 389340, India

M. Rana
Crop Production Division, Indian Council of Agriculture Research-Indian Grass and Fodder Research Institute, Jhansi 284003, Uttar Pradesh, India

Dulal Chandra Roy
Department of ILFC, WBUAFS, Mohanpur, Nadia 741252, West Bengal, India

Shyamashree Roy
Regional Research Station, Old Alluvial Zone, Uttar Banga Krishi Viswavidyalaya, Majhian, Patiram, Dakshin Dinajpur, India

R. P. Sah
Crop Production Division, ICAR-National Rice Research Institute, Cuttack, Odisha, India

Abhijit Saha
Department of Agronomy, College of Agriculture, Lembucherra 799210, Tripura, India

S. Sarkar
Department of Genetics and Plant Breeding, Bidhan Chandra Krishi Viswavidyalaya, Mohanpur 741252, Nadia, West Bengal, India

Savitri Sharma
Department of Agriculture, Jagannath University, Chaksu, Jaipur, India

Minu Singh
Dairy Extension Division, NDRI, Karnal, Haryana 132001, India

O. N. Singh
Crop Production Division, Indian Council of Agriculture Research National Rice Research Institute, Cuttack 753006, Odisha, India

Bireswar Sinha
Department of Plant Pathology, College of Agriculture, CAU, Imphal, India

Mahesh B. Tengli
Division of Dairy Extension, National Dairy Research Institute, Karnal, Haryana 132001, India

B. C. Thangjam
Department of Entomology, College of Agriculture, Tripura, Lembuchhera, West Tripura, Tripura

Parveen Zaman
Pulse & Oilseed Research Sub-station, Beldanga, Murshidabad, West Bengal, India

LIST OF ABBREVIATIONS

ADF	acid detergent fiber
ANF	antinutritional factor
AVKB	Avika Bajra Chari
Bt	*Bacillus thuringiensis*
CF	crude fiber
CP	crude protein
DAS	days after sowing
DM	dry matter
DMY	dry fodder yield
EC	emulsifiable concentrate
EE	ether extract
FYM	farm yard manure
GFB	Gujarat fodder bajra
GFY	green fodder yield
GHG	greenhouse gas
ICRAF	International Council for Research in Agroforestry
ICRISAT	International Crop Research Institute for the Semi Arid Tropic
IGFRI	Indian Grassland and Fodder Research Institute
IPCC	Intergovernmental Panel on Climate Change
IPM	integrated pest management
IVDMD	in vitro dry matter digestibility
NFE	nitrogen free extract
NPV	nuclear polyhedrosis virus
PSB	phosphorus solubilizing bacteria
RDF	recommended dose of fertilizer
SADC	Southern African Development Community
SMIP	Sorghum and Millet Improvement Program
TA	total acid
TMR	total mixed ration
TNAU	Tamil Nadu Agricultural University
WG	wettable granule

ACKNOWLEDGMENTS

First of all, I ascribe all glory to the gracious "Almighty Allah" from whom all blessings come. I would like to thank him for His blessing to write this book.

I express my grateful thanks to my beloved wife, Parveen, for her wholehearted assistance. I express our deep sense of regard to my Abba, Maa, Jiju, Mehebub, Ismat didi, Tuhina, whose provided kind cooperation and constant encouragement.

With a profound and unfading sense of gratitude, I wish to express our sincere thank to the Bidhan Chandra Krishi Viswavidyalaya, India, for providing me with the opportunity and facilities to execute such an exciting project and for supporting my research.

I convey special thanks to my colleagues and other research team members for their support and encouragement and for helping me in every step of the way to accomplish this venture.

I am grateful to Dr. Md. Wasim Siddiqui, Mr. Ashish Kumar, and Mr. Rakesh Kumar from Apple Academic Press for helping me to accomplish my dream of publishing this book series, *Forage Crops of the World.*

PREFACE

Fodder production depends on soil type, land capability, cropping pattern, climate, and socioeconomic conditions. Agricultural animals are normally fed on the fodder available from cultivated areas, supplemented to a small extent by harvested grasses and top feeds. The three major sources of fodder are crop residues, cultivated fodder, and fodder from trees, pastures, and grazing lands. Forage crops are essential for quality milk and meat production. The patterns and types of fodder crops vary as per geographical location. In many countries of the world, people are not paying adequate attention to the feed and fodder for livestock and dairy animals. In addition, green fodder, hay, and silage also are important factors for their health, milk, and meat production during lean periods. Wide forage diversity exists throughout the world. Cultivated land is gradually decreasing, and within that, land resources to meet the need for food and fodder production to feed the world is also decreasing. Moreover, land resource for major fodder production is limited. We have to manage well to grow fodder crops with limited resources. The major fodder crops that are most nutritious to the animals need to be adopted in our cropping systems. Dual purpose crops have to be grown in cropping systems so that food and fodder grown together can meet the demand under sustainable agriculture. Most of the fodder crops also have the human food value. In this respect, best utilization of fodder crops has to be adopted.

PART I
Nonleguminous Forages

CHAPTER 1

OAT (JAI)

JANCY GUPTA* and MINU SINGH

Dairy Extension Division, NDRI, Karnal, Haryana 132001, India

**Corresponding author. E-mail: jancygupta@gmail.com*

ABSTRACT

Oats are popularly used as feed and fodder. Oat ranks sixth in world cereal production. Oat is preferred for hay making. Slowly it has gained importance worldwide as food, feed, and fodder. Its distribution has increased widely due to its adaptability to a broad range of soil types. Oat is originated from Asia Minor. From single cut, 400–450 quintals of green fodder per hectare, and from multicut, 500–550 quintals of green fodder per hectare are produced. Average yield of grains is about 15–20 quintals per hectare. The chemical composition of the green fodder varies with the stage of harvest. Oat is preferred for conservation in the form of hay all over the world. It contains 7–9% crude protein, which can be increased up to 11% by adding nitrogen fertilizers.

1.1 BOTANICAL CLASSIFICATION

Kingdom: Plantae
Order: Poales
Family: Poaceae
Genus: *Avena*
Species: *sativa*
Binomial name: *Avena sativa* (L.)

1.2 COMMON NAME

Oat, *Jai*, *Ganer*, *Ganerji*, *Togi koddi*

1.3 BOTANICAL NAME

Avena sativa (L.)

1.4 INTRODUCTION

Oats are relished by livestock and suitably used as their feed and fodder. Oat is cultivated primarily as a cereal crop. Oat ranks sixth in world cereal production. Oat is preferred for hay making. Slowly it has gained importance worldwide as food, feed, and fodder. Its distribution has increased widely, due to its adaptability to a broad range of soil types. Oat is originated from Asia Minor. The cultivated oats is believed to have originated from *Avena Sterilis* and *Avena byzantina*. Most cultivated oat species is *Avena sativa*, which accounts for 80% of total world area. Rest of the area is cultivated mostly with *Avena byzantine* and only an area is taken under other species (Bimbraw, 2013). The introduction of oat in India dates back to beginning of 19th century. In India, oat is primarily grown in northern, central, and western zones.

1.5 DESCRIPTION

Oat is a significant fodder crop of *rabi* season. Apart from grains Oat is used as straw, for making hay, silage, and green manure. Oat can be cultivated in varied types of agricultural areas. It has high production potential and requires relatively low management. Oat is more frost tolerant than other cereals. It is even well adapted to areas where other cereal crops struggle, such as acidic or water-logged soils (Bimbraw, 2013).

1.6 ORIGIN AND DISTRIBUTION

Oat is originated in Asia Minor. Oat is produced primarily in countries such as European Union, the United States, Russia, Canada, Poland, France,

Finland, Germany, Belarus, China, Ukraine, and the United Kingdom. Oat was introduced in India about in the beginning of 19th century. In India, it is widely cultivated in northern, central, and western zone. It is grown primarily in Uttar Pradesh, Madhya Pradesh, Haryana, Punjab, Himachal Pradesh, Jammu and Kashmir, Uttarakhand, Rajasthan, Maharashtra, West Bengal, Bihar, Gujarat, and Odisha.

1.7 PLANT CHARACTERISTICS

The fodder oats may be erect, drooping or spreading annuals. They attain heights from 1 to 1.5 m and their number varies between 5 and 8 per plant. The leaves are about 25-cm long and 1.5-cm wide. The panicles are lax and effuse. The main axis of the inflorescence has four to six whorls of branches ending in a single apical spikelet. The grain is long, glabrous, or hairy at the apex. The roots are fibrous and penetrate deep in the soil (Mukherjee and Maiti, 2008).

1.8 CLIMATIC REQUIREMENT

Oat is mostly grown in temperate and subtropical areas, in warmer subtropical and Mediterranean climates; oats can be grown in winter as fodder crop. Oat is well adapted to cooler environment. Its optimum growth is attained in sites with 15–25°C temperature in winter with moist conditions. Although, it can tolerate frost up to some extent but its fodder yield and quality is reduced due to hot and dry conditions (Anonymous, 2012). Annual rainfall requirement ranges from 38 to 114 cm but mostly it does not exceed 76 cm.

1.9 SOIL AND ITS PREPARATION

Oat grows best in loam to clay loam soil with adequate drainage. They produce satisfactory yields on heavy or light soils with proper moisture. It can also be grown under moderate, acidic, or saline conditions (Anonymous, 2012). Sufficiently moist soil and pH of 4.5–6.0 is suitable for oat fodder production. It can also tolerate finer soil texture. For land preparation, plow the land three to four times so that the soil becomes free of weeds. Presowing irrigation is also needed for uniform and quick germination (Bimbraw, 2013).

1.10 VARIETIES

Some of the most adapted and widely cultivated varieties of oat for fodder production in India are: UPO 50, UPO 94, UPO 212, OS 6, OS 7, OL 125, JHO 851, RO 19, JHO-822, Oat 17, HFO 114, Palampur 1, HJ 8, SKO 7, JHO 2001-3, JHO 99-2, Harita, JHO 2004, OL 9, JO 1, IGFRI 99-1, Craigs afterlee, N.P.I, N.P. 3, Wild oats, Sativa kent, Filmingold, and Kent (Anonymous, 2009; Anonymous, 2012; Panda, 2014).

1.11 SOWING TIME

Oat can be sown from early October to 1st week of December. Sowing of oat for multicut forage crop is done in October (Anonymous, 2009; Anonymous, 2012).

1.12 SEED AND SEED INOCULATION

The treatment of seed should be done with Vitavax at the rate of 1–2 g per kg of seed to ensure precaution from covered smut disease.

1.13 SEED RATE AND SOWING METHOD

Seed rate for sowing oat for seed crop production is 70–80 kg/ha and for fodder crop cultivation is 90–100 kg/ha. Seeding should be done in rows 20–25-cm apart by the *kera* or *pora*, with broadcast method. Seed depth should not be more than 5 cm. Shallow seedling is possible with high soil moisture and leads to more emergence (Anonymous, 2009; Bimbraw, 2013).

1.14 INTERCROPPING

Oat is successfully cultivated in combination with mustard and berseem (*Egyptian clover*). Seeds of berseem and oat are taken in 1:1 ratio. Pea (*Pisum sativum*), grass pea (*Lathyrus sativus*), and senji (*Melilotus indica*) also make good combination with oat (Anonymous, 2009; Anonymous, 2012).

1.15 CROP MIXTURE

Broadcasting of pea with oats gives higher green as well as dry fodder yield under rainfed conditions. Cultivation mixture of senji and oat provides higher yield of balanced green fodder.

1.16 CROP SEQUENCE

The suitable forage crop sequence involving oats which can be followed are given below:

Sorghum–oat–maize (1 year)
Maize–oat–maize (1 year)
Sudan grass–oat–maize + cowpea (1 year)
Lobia–oat + mustard–maize + cowpea (1 year)
Sorghum + cowpea–oat + lucerne (1 year)

1.17 NUTRIENT MANAGEMENT (MANURES AND FERTILIZERS)

Apply 20–25 tones of farmyard manure before 10–15 days of sowing. The single cut crop requires application of 80 kg N per hectare, 40 kg P per hectare and for multicut crop a dose of 120 kg N per hectare and 40 kg P_2O_5 per hectare is sufficient; here N is given in split doses of 40 kg per hectare which is top-dressed after each cut. For maintaining a balance between N, P, and K, 20 kg K_2O per hectare should be applied as basal. In zinc deficient soils, zinc sulfate may be applied at the rate of 10 kg per hectare at the time of land preparation (Anonymous, 2009; Mukherjee and Maiti, 2008).

1.18 WATER MANAGEMENT

For proper germination presowing irrigation is necessary, followed by three to four irrigations. Irrigate after an interval of a month and after each cut. In case of multiple cuttings, field must be irrigated after each cutting. If soil is dry, first irrigation is given before preparing the seedbed (Anonymous, 2009; Bimbraw, 2013).

1.19 WEED MANAGEMENT

Oats sown for fodder usually require no weeding as it is sown fairly thick. Under rainfed conditions, it is therefore advisable to hoe the crop to reduce water and nutrient uptake by weeds. Usually, one weeding after 3–4 weeks of sowing is enough. A few tall broadleaf weeds, such as ragweed, goose grass, wild mustard, and button weed, can occasionally be a problem, as they complicate harvest and reduce yields. These can be controlled with a modest application of a broadleaf herbicide, such as 2, 4-DB while the weeds are still small. Use of herbicides varies from type of cropping, when in pure stands, preemergence or postemergence herbicides are useful but in legume mixed oat, herbicides should be applied carefully. Benazolin spray @ 0.75 kg a.i./ha with 600–700 L water may be applied in oat + senji mixture, after sowing 2,4-DB @ 0.75 kg a.i./ha may be applied in oat + legume mixture, at 2–3 weeks after sowing (Singh et al., 2011).

1.20 DISEASE AND INSECT PEST MANAGEMENT

Diseases and insect-pest infestation in grains and fodder crops causes huge losses, not only quantitatively but also qualitatively. Some of the major diseases and insect pest, their causal organisms, symptoms, and control measures are given as follows:

1.20.1 PYTHIUM ROOT ROT

It is caused by *Pythium* spp. and is characterized by transparent areas in roots which may later turn reddish-brown, aerial parts turn yellow and stunted growth. It can be controlled by treating seed with Thiram at the rate of 2 g/kg of seed before sowing.

1.20.2 RHIZOCTONIA ROOT ROT

It is caused by *Rhizoctonia* spp. Symptoms, such as appearance of purplish patches in roots and stunted plant with a stiff erect habit is indicative of root rot. Control measures like seed treatment with Thiram at the rate of 2 g/kg of seed is very effective.

1.20.3 LEAF BLOTCH

It is caused by *Helminthosporium avenae*. The distinguishing symbols are one or more brown lesions on the coleoptiles and long, brick red blotches appear on young plants in third and fourth leaf stages. The disease can be controlled by treating seed with Thiram, Bavistin or Vitavex at the rate of 2 g/kg seed before sowing.

1.20.4 CROWN RUST

It is caused by *Puccinia coronata* var. *avenae*. The appearance of orange color lesion on leaf surface arranged in lines covering the entire leaf surface is symptom of disease. For control of the disease seed may be treated with oxycarboxin at the rate of 2.5 g/kg of seed or spray of Zineb at the rate of 0.2% at fortnightly intervals are effective.

1.20.5 STEM RUST

It is caused by *Puccinia graminis f.* sp. *Avenae*. The symptoms of disease are appearance of brown to dark brown and large oblong pustules on culms, leaf sheaths, and blades of plant and black pustules in later stage. Control of disease can be done by treating seed with oxycarboxin at the rate of 2.5 g/kg or by spray of Zineb at the rate of 0.2% at fortnightly intervals.

1.20.6 BIRD CHERRY APHID

It is caused by *Rhopalosiphum maidis*. In this disease the leaves, leaf sheath, and inflorescence are covered with dark green aphid colonies with a slight white covering; molting and distortion of leaf may also occur. For control of the disease, application of Malathion at the rate of 0.05% in the patches of aphid colony is effective.

1.20.7 ROOT-KNOT NEMATODE

It is caused by *Meloidogyne* spp. The diseased plants show symptoms such as: uprooted diseased plants show distinct galls, spindles like shape, beads or club in the roots. The disease can be controlled by crop rotations, fallow, and deep plowing two to three times during summer.

1.20.8 LOOSE SMUT

It is caused by *Ustilago avenae*. All parts of florets are converted into smut sori which are covered by a thin membrane. This membrane ruptures easily on emergence from the boot to expose the smut spores. The disease can be controlled by seed treatment with formalin diluted in equal volume of water, sprinkled and the grains are wetted thoroughly then left covered with moist gunny bags for 19 h.

1.20.9 COVERED SMUT

It is caused by *Ustilago kolleri*. The kernels are replaced by sori and are enclosed in a persistent membrane composed of pericarp and floral bracts which are not damaged. For disease control, seed is treated with formalin diluted in equal volume of water, liquid is sprinkled and the grains are wetted thoroughly then left covered with moist gunny bags for 19 h.

1.21 HARVESTING

For single cut crops harvesting should be done when the crop is at 50% bloom stage. For multiple cuttings, the first cutting should be done at about

70–75 days after seeding and subsequent cutting at an interval of 50–55 days (at the dough stage) (Bimbraw, 2013).

1.22 YIELD

From single cut, 400–450 quintals of green fodder per hectare, and from multicut, 500–550 quintals of green fodder per hectare are produced.

1.23 SEED PRODUCTION

Average yield of grains is about 15–20 quintals per hectare.

1.24 NUTRITIVE VALUES

The chemical composition of the green fodder varies with the stage of harvest. Oat is preferred for conservation in the form of hay all over the world. It contains 7–9% crude protein, which can be increased up to 11% by adding nitrogen fertilizers (Anonymous, 2012).

1.25 UTILIZATION

It is an ideal fodder as a green crop, silage, and hay; mostly it covers the lean periods of year.

1.26 SPECIAL FEATURES TOXICITIES

Oat contains protease inhibitor as antinutritional factor (Anonymous, 2012).

1.27 COMPATIBILITY

Oat as fodder is highly palatable for livestock.

FIGURE 1.1 **(See color insert.)** Oats fodder: before harvesting (Var.—OS-6).

KEYWORDS

- **oats**
- **agronomic management**
- **fodder yield**
- **nutritive value**

REFERENCES

Anonymous. *Handbook of Agriculture*, 6th ed.; Directorate of Information and Publications of Agriculture, ICAR: New Delhi, India, 2009; pp 1054–1372.

Anonymous. *Nutritive Value of Commonly Available Feeds and Fodders in India*; National Dairy Development Board: Anand, India, 2012.

Bimbraw, A. S. *Production, Utilization and Conservation of Forage Crops in India*; Jaya Publishing House: Delhi, India, 2013.

Mukherjee, A. K.; Maiti, S. *Forage Crops Production and Conservation*; Kalyani Publishers: New Delhi, India, 2008; pp 44–86.

Panda, S. C. *Agronomy of Fodder and Forage Crops*; Kalyani Publishers: New Delhi, India, 2014; pp 13–16.

Singh, A. K.; Khan, M. A.; Subash, N.; Singh, K. M. *Forages and Fodder: Indian Perspective*; Daya Publishing House: Delhi, India, 2011.

CHAPTER 2

MAIZE (CORN)

R. P. SAH[1], ANJANI KUMAR[1*], M. RANA[2], and U. KUMAR[1]

[1]*Crop Production Division, ICAR-National Rice Research Institute, Cuttack, Odisha, India*

[2]*Crop Production Division, ICAR-Indian Grass and Fodder Research Institute, Jhansi, Uttar Pradesh, India*

**Corresponding author. E-mail: anjaniias@gmail.com*

ABSTRACT

It is one of the world's third major crops, after rice and wheat. It is a warm season grass that is being used for diverse purposes globally, for example, for human food, ethanol production, biogas, green fodder, feed for livestock, and raw material for producing bioplastics. The green fodder of this crop is widely acceptable by dairy farmers because of its higher digestibility than sorghum, bajra, and other nonleguminous forage crops.

2.1 BOTANICAL CLASSIFICATION

Kingdom: Plantae
Division: Magnoliophyta
Class: Liliopsida
Order: Poales
Family: Poaceae
Genus: *Zea*
Species: *mays*

Based on kernel types, maize has been classified into seven kernel types. Of these, five types (dent, flint, floury, sweet, and waxy) are commercially produced. A brief description of these races based on endosperm and glume characteristics is as follows (Acquaah, 2007).

1. Dent corn (*Zea mays indentata*): It is popularly known as dent corn because of dent is formed on top of the kernel having yellow or white color. Varying amounts of soft starch cause indentations at the tip of the kernel when they dry. Mostly grown in United States.
2. Flint corn (*Zea mays indurata*): This was first developed by Europeans. Rounded top kernels. Mostly grown in Europe, Asia, Central America, and South America. Preferable type for India. Little or no soft starch is found, and the kernels are translucent and shiny.
3. Popcorn (*Zea mays everta*): The cultivation is mainly confined to the new world. The endosperm has only hard starch. The grains are used as popcorn confections.
4. Flour corn (*Zea mays amylacea*): It look like flint corn in appearance and ear characteristics. The grains contain soft starch and little or no dent (hard starch). This is one of the oldest types of maize grown widely in the United States and South Africa.
5. Sweet corn (*Zea mays saccharata*): The sugar and starch are the major component of the endosperm. Immature kernels have sweetish taste and after maturity the kernels become wrinkled. Mostly grown in peri-urban regions of the country.
6. Waxy corn (*Zea mays ceratina*): This type has qualities similar to mochi rice. The kernels contain glutinous type of starch called amylopectin (up to 100%) and looks to have waxy appearance with gummy starch. The origin is supposed to be in China but many waxy hybrids developed in the United States are producing starch similar to that of tapioca.
7. Baby corn: The immature, young ear of female inflorescence of maize is known as baby corn. The young cob harvested before fertilization when the silk has just emerged.

2.2 COMMON NAMES

Maize is also known as corn internationally. In Asian countries it is known as Milho, Yu mai, Makai, etc. Mahiz in France, andmielie (Afrikaans) or mealie in Southern Africa. In India it is known as Makka, Makai, Bhutta (Hindi),

Chujak (Manipuri), Bhutta (Marathi), Makkacholam (Tamil), Makkacholam (Malayalam), Mokkajavanalu (Telugu), Makkejola (Kannada), and Mako (Konkani).

The word maize is derived from the Spanish form of the indigenous Taíno word for the plant, *mahiz*. This crop was domesticated by indigenous peoples in Mexico (the evolution of corn) about 10,000 years ago. In places outside North America, Australia, and New Zealand, *corn* often refers to maize in culinary contexts. This is due to use of additional word for different use of maize, as sweet corn, baby corn, popcorn, etc.

2.3 SCIENTIFIC NAME

Zea mays (L.)

2.4 INTRODUCTION

Maize was named "mais" in Spanish, or "maize," from which Linnaeus coined its Latin name, *Zea mays*. It is one of the world's third major crops, after rice and wheat. It is a warm season grass that is being used for diverse purposes globally, for example, for human food, ethanol production, biogas, green fodder, feed for livestock, and raw material for producing bioplastics. The maize crop is used in the form of green fodder for direct feeding (cut at 50% tasseling stages) and for silage preparation, where plant is cut down at physiological matured stage due to stay-green traits and dry stover at maturity. Maize was, for long time, mostly used as fresh green fodder during lean period (summer) for animals. Plant harvested at pre-flowering stage has low dry matter content than reproductive and grain-filling stages. Forage conservation by ensiling was first explored by Reihlen Stuttgart in the 1860s. Now, maize hybrids produced in different countries have typically 50% grain and 50% forage at physiological maturity of the plant. At this stage, the plant contains usually 70–75% moisture which is good for whole-plant maize silage. Worldwide total stover/straw is estimated to be 3.8 billion metric tons with cereals contributing 74%, sugar crops 10%, legumes 8%, tubers 5%, and oil crops 3% (Lal, 2005). These diversified uses of maize now increase the demand of growing maize as dual purpose, to ensure grain and forage production simultaneously.

Maize (*Z. mays* L.) is a multipurpose crops having wider adaptability under different agroclimatic situations. Due to its high yield potential among other cereals, it is also known as queen of cereals. It is the third most important cereals cultivated in about 160 countries, covering 150 million hectare that contributes 36% (782 million t) in the global grain production. Maize grain contains approximately 72% starch, 10% protein, 4% fat, and 365 Kcal/100 g energy, whereas green fodder contains 10–12% of crude protein, 25–30 crude fiber, and 52–65% in vitro dry matter digestibility (IVDMD). Moreover, dry stover as % dry matter contains crude protein of 3.5–8.8%. The highest maize producing country in the world is the United States, followed by China, Brazil, and Mexico. Argentina and India together produce approximately 517 million metric tons/year (2011). Maize is processed into a variety of food and industrial products such as production of starch, sweeteners, oil, beverages, glue, industrial alcohol, and ethanol.

In India, maize production has increased at ~4.2% compound annual growth rate from ~19.7 million t in 2008–2009 to ~24.3 million t in 2013–2014. The largest use of maize in Asia is for animal feed (around 70% of total volumes). However, less than 20% of corn produced is used for direct consumption. The green fodder is mainly used to feed the milch cattle to boost the milk production to a greater extent. The crop is grown in over 0.9 million ha in India throughout the year. This C4 plant can produce high fodder production capacity in short duration. The green fodder of this crop is widely acceptable by dairy farmers because of its higher digestibility than sorghum, bajra, and other nonleguminous forage crops.

2.5 BOTANICAL DESCRIPTION

The cultivated maize (*Z. mays* L.) plant is erect, single stem, and ranges from 1–4 m in height, with leaf blades ranged from 50- to 90-cm long. The monoecious plant forms a terminal raceme (branched inflorescence) of male flowers (tassel), while the axillary female inflorescences form cylindrical "cobs," each with 16–30 rows of spikelet's (which develop into "kernels" when the seeds mature), and long protruding styles (silk). Mature kernels are typically white or yellow, but different colors also develop through breeding.

The top of the stem ends with tassel, an inflorescence of male flowers. Cob produced at the middle of the plant. Each silk may become pollinated to produce one kernel of corn. Stems are firm, erect, solid, 2–3 m in height with distinct nodes, internodes, casting off flag leaves at every node. The intermodal distance is 20–30 cm. Leaves are simple leaf with entire margin, parallel venation, linear shape, and alternately arranged. The entire plant surface is rough due to impregnation of silica. Lower leaves are longer and broader than upper (50–100-cm long and 5–10-cm wide). Male inflorescence is terminal, staminate, compound raceme or panicle and female inflorescence is axillary, pistillate spadix. Tassel (male inflorescence) consists of many paired spikelets. The lower spikelets are sessile and the upper one is stalked. Each spikelet is two-flowered and four glumes. The ears (female inflorescence) are tightly covered over by 8–13 layers of leaves. The silks (long-protruding styles) are elongated that look like tufts of hair, emerged from cob; at first green and later red or yellow after fertilization. About 5% of the kernels on a cob are produced as a result of self-pollination. The pollen shedding begins earlier than silk emergence (1–3 days before) and continues 3–4 days after the silks are receptive. A single tassel may produce 25 crore pollen grains. The viability of pollen grain is about 8–12 h depending upon the season, temperature, and humidity of the environment. The kernel of corn has a pericarp of the fruit fused with the seed coat, typical of the grasses. An ear produces 200–400 kernels under optimum fertilization. Roots are of adventitious fibrous type. Primary root aborts after germination, and is replaced by fibrous adventitious ones from the base of stem. The important traits that increase the green fodder yield are plant height, leaf number, and stem thickness.

2.6 ORIGIN, EVOLUTION, AND DISTRIBUTION

Human selection has led this crop into cultivated form. Its major center of origin and evolution was in Mexico and Central America. Several theories have been proposed related to origin of maize. The Mesoamerican region located within middle South Mexico and Central America has been recognized as one of the main centers of origin, whereas Randolph (1959) proposed that maize was domesticated, independently, in the Southwestern United States, Mexico, and Central America. In the late 1930s, Paul Mangelsdorf suggested that domesticated maize was the outcome of a series of hybridization events between unknown wild maize and a species of *Tripsacum*, a related genus.

Matsuoka et al. (2002) confirmed that cultivated maize has been derived from hybridization between a small domesticated maize (a slightly changed form of a wild maize) and a grass such as teosinte of section *Luxuriates*, either *Zea luxurians* or *Zea diploperennis*. The centers of diversity occurred in Mexico, Guatemala, the West Indies, Colombia, Peru, Ecuador, and Brazil. One of the most primitive cultivated varieties of corn was Nal-Tel, which resulted in the development of the Maya civilization in the lowlands of Guatemala and the Yucatan peninsula.

2.7 PLANT CHARACTERISTICS

The species *Z. mays* is an allogamous, cross-pollination by anemophily that propagates through seed produced. Maize is a water indicator plant which responds to both water logging and low moisture. The genus *Zea* typically characterized by terminal male inflorescences with paired staminate spikelets and lateral female inflorescences with single or paired pistillate spikelets contains four species as given below: (1) *Z. mays* (2n = 2x = 20)—corn, (2) *Zea mexicana* (2n = 2x = 20)—annual teosinte, (3) *Zea perennis* (2n = 4x = 40)—perennial tetraploid teosinte, and (4) *Z. diploperennis* (2n = 2x = 20)—perennial diploid teosinte.

The wild and domestic diploid *Zea* species, namely, *Z. mays*, *Z. diploperennis*, *Z. mexicana*, and *Z. luxurians* (n = 10) cross freely and their hybrids are fully fertile. However, there is a perennial tetraploid (n = 20) species (*Z. perennis*) which is now extinct in the wild but is still maintained. Maize is generally protandrous, that is, male spikelet's mature earlier than the female spikelet's (Table 2.1).

TABLE 2.1 Growth, Yield and Quality Traits Range.

Traits	Range
Green fodder yield (t/ha)	30–50
Crude protein (%)	4–8
Plant height (cm)	250–320
Number of leaves	13–17
Stem thickness (cm)	1.7–2.2
Leaf–stem ratio	0.2–0.32

2.8 CLIMATIC REQUIREMENT

Maize does well on a wide range of climatic conditions, and it is grown in the tropical as well as temperate regions, from sea levels up to altitudes ranging from 2500 to 3000 m. It is however susceptible to frost at all stages of its growth. Maize is a facultative long-night plant and flowers in a certain number of growing degree days >50°F (8–10°C) in the environment to which it is adapted. The magnitude of the influence that long nights have on the number of days that must pass before maize flowers is genetically prescribed and regulated by the phytochrome system. Photoperiodicity can be eccentric in tropical cultivars, while the long day's characteristic of higher latitudes allows the plants to grow so tall that they do not have enough time to produce seed before being killed by frost.

Maize has C4 metabolism, highly efficient in harvesting the sun's energy and converting it to growth products. It requires considerable moisture and warmth from germination to flowering. The most suitable temperature for germination is 21°C and for growth is 32°C. Extremely high temperature and low humidity during flowering damage the foliage, desiccate the pollen, and interfere with proper pollination. About 50–75 cm of well-distributed rain is conducive to proper growth.

2.9 SOIL AND ITS PREPARATION

Forage maize is grown in almost all types of soil (loamy sand to clay loam), but major efforts may be required to make production feasible. It can be grown best with good drainage and soil aeration. In clayey soils, poor growth of maize plant is visible due to compaction. The optimum soil pH is 5.5–7.5 but pH 6.0–6.5 is most preferred. In this range, calcium and magnesium are optimally available, as are the applied phosphates. A good maize soil should have cation exchange capacity around 20 (mequiv/100 g of soil) and a water-holding capacity of 2 in. per feet of soil. This crop is also known as water indicator crop due to being sensitive to excess moisture stress or water logging. It is desirable to avoid low lying fields having poor drainage. Therefore, the fields having provision of proper drainage should be selected for cultivation of maize. The alluvial soils of Uttar Pradesh, Bihar, and Punjab of India are suitable for raising this crop. Normally, 4–5 harrowing are required before taking up planting with the onset of monsoon.

2.10 VARIETIES

2.10.1 TYPES OF VARIETIES

Late-maturing varieties—mature in 110–120 days, suitable for areas with a long rainy season. Intermediate-maturing varieties—mature in 100–110 days, early-maturing varieties—mature in 90–100 days, can be planted after intermediate-maturing varieties. Extra-early varieties—mature in 80–90 days and can be planted in areas with a very short rainy season.

Some famous varieties grown in different countries were Dea in Austria; LG2080 in Belgium; Antarès, Apache, Aviso, Dea, DK250, Mona, Fanion, Keo in France; DK250, Golda, Helix, Mona in Germany; Lorena, Ranger, Fedro in Italy; Brutus, LG2080, Splenda in Netherlands; and LG2080 in United Kingdom.

2.10.2 SOME IMPORTANT INDIAN MAIZE VARIETIES

African tall composite: It is a composite of seven genotypes (H-611 C, H-611, H-611 (R)C3, K-III × EC-573 (R12) C3, Ukiri Comp A (F) C5 × Ukiri Comp A, (F) C3, Chitedge Comp A and Ilonga Comp) developed through modified mass selection technique. This cultivar has high dry matter and crude protein content and more seed yield potential than other grain varieties. The variety is resistant to foliar diseases and stem borer. The average plant heights 260 cm and provides 60–70 t/ha green fodder and 3 t/ha grain.

APFM-8: It is suitable for cultivation in south zone of the India. This is a synthetic variety derived from Varun (V-41) and Palampur local varietal cross advanced by mass selection. It has high green biomass , nonlodging, orange grain variety with plant height of 180–200 cm, and a sturdy plant. The seed to seed maturity is 90–95 days during wet season and 105–110 days in winter. It provides 35 t/ha green fodder and 7.5 t/ha dry fodder.

J-1006: Developed by crossing "Makki safed 1-DR" × "Turpeno PB." It is resistant to maydis blight, brown striped downy mildew, and stem borer.

Pratap Makka Chari 6: Developed by compositing 11 early to medium white-seeded entries. It is a medium tall variety with relatively low ear placement. Its stem is strong, medium thick, and resists lodging. The ears are long, thick with a tight husk cover. It matures in 90–95 days. The green fodder yield potential is 45–50 t/ha.

2.11 SOWING TIME

Maize can be grown in all seasons; monsoon, postmonsoon, winter, and spring. During rabi and spring seasons to achieve higher yield at farmer's field, assured irrigation facilities are required. During *kharif* season, it is desirable to complete the sowing operation 12–15 days before the onset of monsoon. However, in rainfed areas, the sowing time should coincide with onset of monsoon. The optimum time of sowing is given below (Table 2.2).

TABLE 2.2 Season, Sowing Time, and Condition.

Season	Optimum time of sowing	Condition
Wet season (monsoon)	Last week of June to first fortnight July	Grown under rainfed condition
Dry season (winter)	Last week of October for intercropping and up to 15th of November for sole crop	Grown under completely irrigated condition
Spring (summer)	First week of February	Grown under completely irrigated condition

2.12 SEED TREATMENT

Seed treatment aims to protect seedlings against soil and seed-borne diseases and pests, to enhance germination and to promote seedling emergence during the critical first few weeks after planting. To control seed-borne pathogens of fungal disease, treat the seed with carbendazim or Thiram at 2 g/kg seed (Table 2.3).

TABLE 2.3 Seed Treatment with Fungicides and Insecticides.

Disease/insect-pest	Fungicide/pesticide	Rate of application
Turcicum leaf blight (TLB), banded leaf and sheath blight (BLSB), maydis leaf blight (MLB)	One part carbendazim 50% WP + one part Captan 50% WP	2.0 g/kg of seed
Pythium stalk rot	Captan 50% WP	2.5 g/kg of seed
Termite and shoot fly	Imidachloprid (600 g a.i./kg imidaclоprid)	35 mL/kg of seed
Biofertilizer	Azospirillum	10–12 mL/kg of seed

2.13 SEED RATE AND SOWING METHOD

The seed rate may be decided as per purpose for which maize was grown. Generally, 30-cm row to row and 10-cm plant to plant is recommended for fodder maize. Sowing in rows is generally done with drill or by dropping the seed behind the plow. To achieve higher productivity and to trap higher resource-use efficiencies, optimum plant population should be maintained. The following crop geometry and seed rate should be adopted (Table 2.4).

TABLE 2.4 Purpose, Seed Rate, and Spacing.

Purpose	Seed rate (kg/ha)	Row to row × plant to plant	Recommended in
Fodder	50–75	30 × 10	All season
Green cob (normal maize)	20	75 × 20	Wet season
		60 × 20	Dry or summer season
Grain (normal and Quality Protein Maize (QPM))	20	75 × 20	Wet season
		60 × 20	Dry or summer season
Sweet corn (immature cob)	8	75 × 25	All season
Baby corn (unfertilized ear)	25	60 × 20	Wet season
		60 × 15	Dry or summer season
Popcorn (dried grain)	12	60 × 20	All season

2.14 MAIZE-BASED CROPPING PATTERN

Maize has wide adaptability and compatibility under diverse soil and climatic conditions, and hence it is cultivated in sequence with different crops under various agroecologies of the country. Among different maize-based cropping systems, maize–wheat ranks first having 1.8 million hectare area mainly concentrated in rainfed ecologies. Maize–wheat is the third most important cropping systems after rice–wheat and rice–rice that contributes about 3% in the national food basket. Recently, due to changing scenario of natural resource base, rice–maize has emerged as a potential maize-based cropping system in peninsular and eastern India. In peri-urban interface, maize-based high-value intercropping systems are also gaining importance due to market-driven farming. Further, maize has compatibility with several crops of different growth habit that led to development of various intercropping systems (Table 2.5).

TABLE 2.5 Maize-based Sequential Cropping Pattern for Different Ecologies.

Irrigated	Rainfed
Fodder maize–potato–wheat	Fodder maize–mustard
Fodder maize–mustard	Fodder maize–legumes
Summer rice–fodder maize–mustard	Fodder maize–barley
Fodder maize–maize	Fodder maize–safflower
Fodder maize–legumes	Rice–potato–fodder maize
Fodder maize–early potato–wheat–mung bean	Sorghum–fodder maize
Fodder maize–wheat–mung bean	Fodder maize–sorghum–pulses
Fodder maize–potato–sunflower	Fodder maize–maize–pearl millet
Rice–potato–fodder maize	Rice–fodder maize + cowpea
Fodder maize–potato–onion	
Mung bean–fodder maize–toria–wheat	
Fodder maize–groundnut–vegetables	
Rice–fodder maize–pearl millet	
Rice–rice–fodder maize	
Maize–chickpea	

2.15 NUTRIENT MANAGEMENT

Maize is highly responsive to available nutrient in soil. It stimulates vegetative growth, regulates protein content in plant, and intensifies green color. The deficiency symptoms include pale green or yellowish leaves, premature drying of leaf tip and midribs, stunted growth, and lower productivity. For green fodder cultivation the rate of application of nitrogen (N) is 80 kg N/ha, which should be applied in two split doses. First dose 40 kg/ha N at the time of sowing as basal dose and second dose 40 kg/ha after 30 days after sowing (DAS).

Maize critically requires phosphorous (P) in early stage of plant, when the root system very less develops to take up P from soil. P deficiency symptoms appear as purplish leaves, stunted plant growth, delayed flowering, and improper root development. For green fodder cultivation, the rate of application is 40 kg P_2O_5/ha at sowing time as basal dose.

Zinc (Zn) deficiency is sometimes observed in crop which affects the forage and grain quality. ZnSO4 @ 20 kg ha^{-1} as basal fertilizer may be applied to soil. Overall, plant may also show other micronutrient deficiencies if soil lack the micronutrients. For which micronutrients can be applied after sowing the seeds.

2.16 IRRIGATION MANAGEMENT

The irrigation water management depends on season as about 80% of maize is cultivated during monsoon season, particularly under rainfed conditions. However, in areas with assured irrigation facilities, irrigation should be applied as and when required by the crop, and first irrigation should be applied very carefully wherein water should not overflow on the ridges/beds. In general, the irrigation should be applied in furrows up to two-third height of the ridges/beds. Young seedlings, knee high stage, flowering, and grain filling are the most sensitive stages for water stress, and hence irrigation should be ensured at these stages. In raised bed planting system and limited irrigation water availability conditions, the irrigation water can also be applied in alternate furrow to save more irrigation water. In rainfed areas, tied ridges are helpful in conserving the rainwater for its availability in the root zone for longer period. For winter maize, it is advisable to keep soil wet (frequent and mild irrigation) during 15 December to 15 February to protect the crop from frost injury.

2.17 WEED MANAGEMENT

Weeds are the serious problem in maize, particularly during wet/monsoon season they compete with maize for nutrient and may cause yield loss up to 35%.

The common species of weeds are:

(a) Monocot weeds: *Cenchrus echinatus*, crabgrass (*Digitaria* spp.), wiregrass (*Eleusine indica*), and foxtail (*Setaria verticillata*) of family Gramineae. Nutgrass (*Cyperus rotundus*) of family of sedges, the Cyperaceae (tolerance of most common herbicides), etc. The "nuts" (tubers) of this sedge are long-lived and thousands in number. The effective removal of the nutsedges is possible by use of methyl halosulfuron herbicide which has been formulated to control weeds without harming grasses.

(b) Broad-leaved weeds: Amaranths (*Amaranthus* spp.), purslane (*Portulaca oleracea*), honohono grass (*Commelina diffusa*), the spurges (*Euphorbia hirta*), etc.

Long-term field management can minimize weeds by periodically tilling or killing perennial weeds and by preventing annual weeds from forming seeds. Crop rotation with legume cover crops such as *Crotolaria juncea* (sunn hemp), *Canavalia ensiformis*, or *Mucuna* spp. is useful depending upon the time availability between two crops.

During the initial stage, the growth of the maize plants is suppressed by weeds. Weeding is generally done between the rows by hand weeder, whereas the weeding within the row is done by hand. Two or three weeding is necessary for good production from the crop. After first weeding (30–45 DAS) the crop is earthed up to provide for better standability. No intercultivation after flowering is necessary, as it is likely to damage the lateral roots. For the fodder crop of maize, less of weeding is needed, since the soil surface is nearly covered by a dense population of the maize plants. Only first weeding after 30–45 DAS is compulsory for fodder maize.

Herbicide application may be done for effective control over weeds. Herbicides used for control of growing weeds before planting maize are either systemic herbicides, effective against most types of weeds, or contact herbicides, effective against most annual weeds. The effectiveness of preemergence herbicides also depends on field moisture and the avoidance of clumping of soil particles in upper soil layers. Use of mixtures of two or more herbicides often gives better results and prevents domination by a single weed species. Postemergence herbicides, applied during maize growth, may be recommended if needed to control weeds when preplant or preemergence application is inadequate.

Atrazine, a selective herbicide is used to restrict the emergence of wide spectrum of weeds. Pre-emergence application of atrazine @ of 1.0–1.5 kg a.i. ha^{-1} in 600 L water, Alachlor @ 2–2.5 kg a.i. ha^{-1}, metolachlor @ 1.5–2.0 kg a.i. ha^{-1}, and pendimethalin @ 1–1.5 kg a.i. ha^{-1} is effective way to control annual and broad-leaved weeds. Under heavy weed infestation, post emergence application of paraquat can also be done (as protected spray using hoods) (Table 2.6).

TABLE 2.6 List of the Pre- and Postemergence Herbicides Used in Maize.

Type, subtype	Chemical names	Target
Preemergence (applied to the soil to prevent weed emergence)		
Thiocarbamate	EPTC, butylate	Persistent grasses and the nutgrasses
Triazine	Atrazine, simazine, cyanazine	Widely used on corn, applied before or after seedling emergence. Effective against broad-leaved plants but less effective against grassy weeds
Acetanilide	Alachlor, propachlor, metalachlor, dicamba	Common grasses
Dinitroanalines	Pendimethalin	Used as both pre and postemergence in maize field
Postemergence (applied to growing weeds)		
Systemic	Glyphosate, glufosinate	All vegetation
Contact	Paraquat, methyl halosulfuron	Paraquat can be used if directed carefully to weeds that are growing in the inter-rows and not allowed to contact the corn
Plant hormone	2,4-D	Control of broad-leaved weeds

2.17.1 PRECAUTION

- While spraying, move backward so that the atrazine film on the soil surface should not be disturbed.
- Use three boom flat-fan nozzles for proper ground coverage and saving time.
- Under zero tillage condition, preplant application (15–25 days prior to seeding) of nonselective herbicides, namely, glyphosate @ 1.0 kg a.i. ha^{-1} in 400–600 L water or paraquat @ 0.5 kg a.i. ha^{-1} in 600 L water is useful to control the weeds.
- Ecologically sound practices include crop rotation, use of live mulches, and other nonchemical practices that minimize weeds in sustainable agricultural systems.

2.18 DISEASE AND INSECT-PEST MANAGEMENT

Insect and disease management: if they are not managed at proper time, they lead to yield loss. The loss due to major diseases of maize in India is about 13.2%. Of which, foliar diseases contribute a loss of 5% and stalk rots, root rots, ear rots is 5%. The major insect and diseases and their management practices are given below (Table 2.7).

TABLE 2.7 Insect-pest and Diseases Management.

Insect/pest and diseases	Symptoms	Control
Insects		
Stem borer or stalk borer (*Chilo partellus*)	Occurs during monsoon season, lays eggs 10–25 days after germination on lower side of the leaves, enters in the whorl and cause damage in the leaves	Foliar spray of 0.1% endosulfan (35 EC) (700 mL in 250 L water) 10 days after germination Release of 8 trichocards (*Trichogramma chilonis*) per hectare at 10 days after germination Intercropping of maize with cowpea
Pink borer (*Sesamia inference*)	Mostly cause damage in winter season, moth is nocturnal and lays eggs on lower leaf sheath, larvae enter the plant near the base and cause damage to stem	Foliar spray of 0.1% endosulfan (35 EC) (700 mL in 250 L water) 10 days after germination
Shoot fly (*Atherigona* sp.)	Appears mostly in spring and summer, attack at seedling stage, they cut the growing point or central shoot which results in to dead heart formation	Seed treatment with imidacloprid @ 6 mL/kg seed (spring maize)
Termites (*Odontotermes obesus*)		Apply fipronil granules in soil @ 20 kg ha^{-1} followed by light irrigation Fipronil @ 2–3 granuled/plant
Disease		
Turcicum leaf blight (*Exserohilum turcicum*)	Prevalent in cooler and high humidity region, long, elliptical, grayish green or tan lesions (2.5–15 cm) appear on lower leaves and progresses upward	Grow tolerant varieties Sprays Mancozeb @ 2.5 g/L at 8–10 days interval
Maydis leaf blight (*Drechslera maydis*)	Appears when warm humid temperate to tropical climate in the cropping period, lesions on the leaves elongated between the veins, tan with buff to brown (or dark reddish brown borders)	Sprays of Mancozeb or zineb @ 2.5 g/L of water

TABLE 2.7 *(Continued)*

Insect/pest and diseases	Symptoms	Control
Common rust (*Puccinia sorghi*)	Appears in subtropical temperate and high land environment, circular to elongate, golden brown to cinnamon brown pustules are visible over both leaf surfaces changing to brownish black at plant maturity	Spray of Mancozeb@ 2.5 g/L of water at first appearance of pustules Growing early maturing variety is preferential
Banded leaf and sheath blight (*Rhizoctonia solani f.* sp. *Sasakii*)	Appears in hot humid foothill region, white lesions (concentric bands and rings on lower leaves and sheaths) on leaves and sheath. Purplish or brown horizontal bands present on white lesions, later on spread to the ears	Stripping of lower 2–3 leaves along with their sheath Grow tolerance varieties Seed treatment with peat-based formulation @ 16 g/kg of *Pseudomonas fluorescens* Soil application @ 7 g/L of water, carbendazim, thiophanate-methyl Foliar spray (30–40 days old crop) of tolclofos-methyl @ 1 g/L or validamycin @ 2.7 mL/L of water

The chemical management in green fodder maize production is not necessary. Farmers may opt for alternate and biological means of management to fodder production. However, for dual-purpose crops and seed production combine use of management strategies would be effective for highest production from the crop.

2.19 HARVESTING AND YIELD

Maize grown for fodder should be harvested at 50% flowering stage. However, for dual-purpose use the cutting time may be different as given below (Table 2.8).

2.20 SEED PRODUCTION OF MAIZE FODDER

After growing the crop with recommended practice, the following additional operations must be done in seed production plots.

TABLE 2.8 Purpose, Yield, Cutting Time and Uses of Fodder Maize.

Purpose	Yield	Cutting time	t/ha	Major use
Fodder	Green fodder	50% flowering	40–50	Green fodder
Green cob (normal maize)	Green cob + green fodder	Green cob stage	5–10 + 25–27	Green cob
Grain (normal and QPM)	Grain + stover/green fodder (stay green)	Physiological maturity	8–11 + 8–10/15–20	Grain
Sweet corn (immature cob)	Immature grain + green fodder	18–22 days after pollination (dried silk)	7–9 + 25–30	Sweet corn
Baby corn (unfertilized ear)	Unfertilized ear (baby corn) + green fodder	Silk comes out 2.0–3.0 cm from the top of ears	5–11 (husked) + 28–30	Baby corn
Pop corn (dried grain)	Matured grain + stover	Physiological maturity	2.5–3 + 3–5	Grain

2.20.1 ROGUING

It helps in maintaining the genetic purity of seeds. During rouging, off type and diseased plants are removed from the seed field. Generally, rouging is done three times in maize and may be extended depending upon the necessity.

- First rouging: during vegetative stage, based on the height of the plant, plant growth, color of petiole, and color of leaf.
- Second rouging: during flowering stage, based on color characteristics of tassel and silk. Dissimilar plants should be removed.
- Third rouging: before harvest, based on color of seed, stone color, and cob characteristics.

2.20.2 ISOLATION DISTANCE

- Since maize is a cross-pollinated crop, it is necessary to isolate seed field from other maize fields.
- At least 400–500-m distance is required to avoid any contamination.

2.20.3 STAGES OF CROP INSPECTION

- At the time of sowing: to monitor the land, isolation distance, planting ratio of male:female, proper sowing time, seed treatment.
- During preflowering/vegetative stage: to verify the rouging and removal of off type plants.
- During flowering stage: to check disease and pest infestation.
- During post-flowering and preharvest stage: to remove the late and diseased plants.
- Differential type of tassel/silk plants.
- Harvesting time: to see the proper time of harvesting.

2.20.4 TIME OF HARVEST

- Harvest the crop when the outer cover of the cob turns from green to white.
- At the time of harvest, moisture content of the seed will be around 25%.

2.20.5 COB SORTING AND SHELLING

- Remove very small sized cobs, having large difference in seed rows and seed color. The cobs with very less seeds also should be removed.
- After drying the cobs (up to grain moisture content at 15%) seeds can also be separated from cob by beating with sticks.
- For grading quality, grade the seeds with 18/64″ round perforated sieves can be used.

2.20.6 SPECIAL OPERATION REQUIRED FOR HYBRID SEED PRODUCTION

2.20.6.1 MALE:FEMALE RATIO

- The male:female ratio depends on pollen shedding potential and synchrony of male and female parents. Male pollen dehiscence should coincide with female silking. The male:female ratio should be 1:2 or 1:3 or 1:4. Even farmers can go for 1:5–1:6 ratios if they go for paired male rows.

2.20.6.2 DETASSELING

- Removal of tassel from the female parent is known as detasseling. It should be done before anthesis. The removed tassel should be thrown far away from the field.

2.21 NUTRITIVE VALUE

Fodder maize is an excellent crop in terms of biomass production and quality. Maize straw is used as animal fodder since the ancient times. For dairy farmers, the green fodder of maize as animal feeding reduces cost of nutrition because its quality is much better than sorghum and pearl millet (possesses antinutritional quality such as hydrogen cyanide in sorghum and oxalate in pearl millet). Further, the dual-purpose use of this crop such as baby corn, sweet corn, etc. proves farmers nutritional security for animals and humans.

Adequate nutrition to animals is essential for high rates of gain, ample milk production, efficient reproduction, and overall profits. However, forage quality in different maize varieties and nutritional requirement in different animal species or classes varies. Hence, analyzing forages for nutrient content will help to guide proper ration (Table 2.9).

TABLE 2.9 Detailed Compositional Analysis of Maize Fodder.

Traits	Stover (% dry matter basis)	Green fodder	Silage*
Dry matter yield		21–23%	22–25%
Crude protein	4–7%	4–8%	7–9%
Crude fiber	32–36%	24–28%	23–25%
Neutral detergent fiber	62–65%	64–70%	65–66%
Acid detergent fiber	42–44%	40–45%	35–37%
IVDMD	35–50%	52–68%	65–67%
Cellulose	31–40%	28–30%	22–25%
Hemicelluloses	20–35%	30–32%	24–285
Total lignin	15–25%	5–6%	4–6%
Ash	4–7%	8–9%	5–6%

IVDMD, in vitro dry matter digestibility.

*Cut at preheading stage.

Information was collected from Hancock (2009) (http://www.georgia-forages.com/questions/040%20FAQ-crop%20residues.pdf); Khan et al. (2011); Ming-yuan (2015); Gupta et al. (1988); Das et al. (2015); and Kamalak et al. (2003), http://corn.agronomy.wisc.edu/Silage/S006.aspx/ Thursday (accessed Jan 19, 2017). The information may vary with varieties and location of the crop grown.

The quality of the maize fodder may be affected by the following.

2.21.1 MATURITY STAGE

Quality declines with advancing maturity (after milking stage forage quality declines). Matured plant has more fibers, forage intake drops drastically by animals.

2.21.2 LEAF TO STEM RATIO

Reduce leaf to stem ratio declines the forage quality. Leaves are more digestible than stems.

2.21.3 VARIETAL EFFECT

Quality of the forage is varying from variety to variety in maize. Fodder maize variety such as J-1006 has high crude protein content.

2.21.4 DRY MATTER DIGESTIBILITY

IVDMD measuring the nutritional quality of forages.

2.22 UTILIZATION

2.22.1 GREEN FODDER

Maize is palatable green forage for animal and livestock. In quality aspect it has high crude protein (CP)%, crude fiber (CF)% minerals, and vitamins.

2.22.2 SILAGE

Silage is the product in which cut forage of high moisture content is fermented to produce a feed that resists further breakdown in anaerobic storage. A good silage has a pH less than 5.0, total nitrogen is less than 15%, lactic acid 50% of the total organic acids, and butyric acid <0.5% of the total dry matter. It has sweet smell, and cattle, goats, and sheep will readily eat it. Good source to prove fodder in lean period.

2.22.3 STOVER

Maize stover (9–12% moisture content) is the portion leftover after the cob is harvested. This consists of the leaves and stalks of maize plants left in a field. It makes up half of the yield of a crop. Stover is widely used as the major source of animal feed in different countries. The ratio of maize stover to grain is assumed to be 1:1. The nutritional quality of maize stover is poor.

FIGURE 2.1 **(See color insert.)** Fodder maize.

KEYWORDS

- **maize**
- **cultivation practices**
- **nutritive value**
- **utilization**

REFERENCES

Acquaah, G. *Principles of Plant Genetics and Breeding*; Blackwell Publishing: United Kingdom. 2007.

Das, L. K.; Kundu, S. S.; Kumar, D.; Datt, C. Fractionation of Carbohydrate and Protein Content of Some Forage Feeds of Ruminants for Nutritive Evaluation. *Vet. World* **2015**. www.veterinaryworld.org/Vol.8/February-2015/12.pdf.

Gupta, H. O.; Singh, J.; Jain, O. P. Brown Midrib Colour as an Index of Forage Quality in Maize (*Zea mays* L.). *Proc. Indian Nat. Sci. Acad.* **1988,** *54*, 2–3. (175–178).

Hancock, D. W. What is the Nutritional Value of Corn Fodder/Stover (Stalks, Shucks, and Leftover Grain after Harvesting)? What is it Worth? The University of Georgia College of Agricultural and Environmental Sciences and the U.S. Department of Agriculture Cooperating, 2009.

Kamalak, A.; Erol, A.; Gurbuz, Y.; Ozay, O.; Canbolat, O.; Tumer, R. Evaluation of Dry Matter Yield, Chemical Composition and In Vitro Dry Matter Digestibility of Silage from Different Maize Hybrids. *Livestock Res. Rural Dev.* **2003,** *15*, 11.

Lal, R. World Crop Residues Production and Implications of Its Use as a Biofuel. *Environ. Int.* **2005,** *31*, 575–584.

Liang, M.; Wang, G.; Liang, W.; ShiPeng, F.; Jing, D.; Peng, S.; Chun-sheng, H. Yield and Quality of Maize Stover: Variation Among Cultivars and Effects of N Fertilization. *J. Integr. Agriculture*. **2015,** *15*, 61077–61082. DOI: 10.1016/S2095-3119.

Matsuoka, Y. Y.; Vigouroux, M. M.; Goodman. A Single Domestication for Maize Shown by Multilocus Microsatellite Genotyping. *Proc. Natl. Acad. Sci.* **2002,** *99* (9), 6080–6084.

Randolph, L. F. The Origin of Maize. *Indian J. Genet. Plant Breed* **1959,** *19*, 1–12.

CHAPTER 3

PEARL MILLET (BAJRA)

SHAKTI KHAJURIA*, A. K. RAI, B. S. KHADDA, RAJ KUMAR, and J. K. JADAV

ICAR-CIAH, KVK-Panchmahal, Vejalpur, Godhra 389340, India

**Corresponding author. E-mail: shaktikhajuria@gmail.com*

ABSTRACT

Pearl millet is an annual grass belonging to the family Poaceae. Pearl millet is a valuable feed and fodder for livestock. Livestock provides draught power, rural transport, manure, fuel, milk, and meat. Most often, livestock is the only source of cash income for subsistence farms and also serves as insurance in the event of crop failure. Further, global energy crisis will lead to utilization of livestock-based bioenergy as well as waste recycling for organic manure and organic forage production for quality animal products. The yield of bajra with improved cultural practices is nearly 30–35 quintals of grain and about 100 quintals of dry stover from a hectare of crop under irrigated conditions. It is high in protein and energy and low in fiber and lignin concentration. Crude protein can range from 9% to 11% in unfertilized soils to 14% to 15% under nitrogen-fertilized conditions. Bajra has a high potential for accumulating toxic levels of nitrate. It is best to avoid grazing younger plants and to avoid overgrazing. Droughty or cold weather can stress plants and increase nitrate levels.

3.1 BOTANICAL CLASSIFICATION

Kingdom: Plantae
Order: Poales

Family: Poaceae
Subfamily: Panicoideae
Genus: *Pennisetum*
Species: *glaucum*

3.2 COMMON NAMES

Pearl millet (English), bajra (Hindi, Urdu, and Punjabi), sajje (Kannada), kambu (Tamil), kambam (Malayalam), sajjalu (Telugu), and bajri (Rajasthani, Gujarati, and Marathi).

3.3 BOTANICAL NAME

Pennisetum glaucum (L.) *R. Br.*

3.4 SYNONYMS

Pennisetum americanum (L.) *Leeke, Pennisetum typhoides* (*Burm. f.*) *Stapf and CE Hubb, Pennisetum typhoideum Rich., Pennisetum spicatum* (L.) *Körn., Setaria glauca* (L.) *P. Beauv* (USDA, 2009).

3.5 INTRODUCTION

Pearl millet [*Pennisetum glaucum* (L.)] is the fourth most important gain crop next to rice, wheat, and sorghum. It is becoming an increasingly important forage crop in many regions of the world. Also, it is a suitable crop for semiarid areas, due to its high resistance to drought stress (Tabosa et al., 1999). Its nutritious grain forms the important component of human diet and stover forms the principal maintenance ration for ruminant livestock during the dry season. In addition to grain and forage uses, pearl millet crop residues are used as fodder, building material, and fuel for cooking, particularly in dry land areas. Pearl millet crop has wide adaptability to local environments. It is a hardy crop and can be grown in areas which are very hot and dry and on soils too poor for crops such as maize and sorghum. Its green forage (without prussic acid, a poisoning potential commonly found in sorghum and Sudan grass) is a valuable feed for livestock. Livestock provides draught power, rural transport, manure, fuel, milk, and meat. Most often, livestock is the only source of cash income for subsistence farms and also serves as insurance in the event of crop failure. Further, global energy crisis will lead to utilization of livestock-based bioenergy as well as waste recycling for organic manure and organic forage production for quality animal products. A majority of the subsistence farmers who typically cultivate this crop are unable to take advantage of high yield potential of pearl millet because of the lack of application of improved management practices. Pearl millet productivity can be increased by growing varieties/hybrids with improved tolerance to drought, resistance to diseases, and with fertilizer applications and soil moisture management.

3.6 PLANT DESCRIPTION

Pearl millet is an annual grass in the family Poaceae. It is an erect annual grass, reaching up to 4-m high with a profuse root system. Culms are slender, 1–3 cm wide. Leaves are alternate, simple, blade linear, pubescent, and minutely serrated, up to 1.5 m long × 8 cm wide. The inflorescence is a panicle, 12–30 cm long. Fruits are grains whose shape differs according to cultivars. It uses C4 carbon fixation (Andrews and Kumar, 1992). Pearl millet grain is considered as a staple food in Africa and India, where it is used to make flour and other foodstuffs. As a feedstuff, it is mainly grown to produce hay, silage, green chop, pasture, and stand over feed grazed directly.

Pearl millet stover is the part of the plant that remains after grain harvest and is a fibrous by-product with a low nutritive value. However, dual-purpose (grain and forage) varieties with higher forage quality are being developed. Pearl millet "brown midrib" mutants have been used to increase forage quality. They give lower yields but contain less lignin, more crude protein, have higher dry matter (DM) degradability and digestibility, and their overall quality does not drop as quickly as they mature, as occurs with normal midrib types (Hassanat, 2007).

Hybrids of pearl millet and Napier grass (*Pennisetum purpureum*) have been developed. They benefit from the desirable characteristics of pearl millet such as vigor, drought resistance, disease tolerance, forage quality, and seed size, whereas Napier grass provides rusticity, aggressiveness, perennity, palatability, and high dry matter yield (Timbo et al., 2010).

3.7 ORIGIN AND DISTRIBUTION

Pearl millet originated in tropical Western Africa some 4000 years ago. The greatest numbers of both wild and cultivated forms of this species occur in this region. From there, it differentiated into *globosum* race and moved to the western side, and it also differentiated into the *typhoides* race that reached Eastern Africa and spread to India and Southern Africa some 2000–3000 years ago. The evolution of pearl millet under the pressures of drought and high temperatures imparted the ability to tolerate drought, nutrient deprived soil, and high temperatures of Indian and African hot deserts more effectively than other cereals such as wheat and rice.

The important pearl millet-growing countries are India, China, Nigeria, Pakistan, Sudan, Egypt, and Arabia. India is the largest producer of pearl millet in the world. Principal pearl millet-growing states are Rajasthan, Maharashtra, Gujarat, Western Uttar Pradesh, Haryana, and Karnataka which accounts for 90% of the total area and 86% of production.

It thrives well where other cereals (maize and sorghum) cannot grow because of drought or heat. It can be found in regions where annual rainfalls range from 125 to 900 mm. Ideal growth temperatures range from 21°C to 35°C. Pearl millet is known to tolerate acid sandy soils and is able to grow on saline soils.

3.8 PLANT CHARACTERISTICS

Pearl millet is grown largely for its ability to produce grain under hot, dry conditions on infertile soils of low water-holding capacity, where other crops generally fail completely.

3.8.1 MATURE PLANT

Pearl millet may grow from 50 cm to 4 m tall, and may tiller profusely under favorable weather conditions.

3.8.2 STEMS

Stems are pithy, tiller freely, and produce an inflorescence with a dense spike-like panicle.

3.8.3 LEAVES

Pearl millet has long leaves that are slender and smooth or have hairy surfaces. The leaves may vary in color, from light yellowish green to deep purple. The leaves are long pointed with a finely serrated margin.

3.8.4 FLOWERS

Pearl millet usually flowers from 40 to 55 days. The flowering structure (inflorescence) in pearl millet is called a panicle or head. The mature panicle is brownish in color.

3.8.5 SEED

The seed begins developing after fertilization and matures 25–30 days later. The seeds are nearly white, yellow, brown, grey, slate blue, or purple in color. The size of the seed is about one-third that of sorghum and the weight about 8 mg on average.

3.8.6 ESSENTIAL PARTS

The grains are an essential part in pearl millet, while the entire crop is used as fodder.

3.9 CLIMATIC REQUIREMENTS

Pearl millet can grow in a wide range of ecological conditions and yield reasonably well even under unfavorable conditions of drought stress and high temperatures. It is mostly grown in countries with hot and dry weather, quite characteristic of the arid and semiarid environments. Bajra is a crop of warm and dry climatic conditions. The ideal temperature for its growth is between 25°C and 31°C (10–20°C at the time of sowing). It is more tolerant to higher temperatures than probably any other major cultivated cereal. Pearl millet hybrids having good seed set at air temperatures as high as 46°C are cultivated during the summer in parts of India.

These useful characteristics mean that pearl millet is finding a new niche in some unexpected places. The best temperature for the germination of pearl millet seed is 23–32°C. Poor emergence and seedling growth may result, if planted before soil temperature reaches 23°C.

The optimum rainfall requirement of pearl millet ranges between 600 and 800 mm. But, pearl millet can be grown in areas which receive even less than 350 mm of seasonal rainfall. Prolonged spells of warm and dry weather are detrimental to the crop, leading to reduced crop yields. At harvest time, dry and warm weather is most suitable. Although pearl millet can respond to good moisture during its growth, it is nevertheless one of the toughest, drought-tolerant crops. It maintains its popularity in the regions where the weather is very unpredictable. Bajra crop is sown between May and September and harvested between October and March. It is grown either as a pure or mixed crop with cotton, jowar, or ragi.

3.10 SOIL AND ITS PREPARATION

Pearl millet can be grown in different soils. It yields best on fertile, well-drained loamy soils. However, it can also grow in shallow soils and in soils with clay, clay loam, and sandy loam texture. Pearl millet does not grow well in soils prone to waterlogged conditions. The crop needs very fine tilth because the seeds are too small. Two to three harrowing and a plowing is followed so that a fine tilth may be obtained to facilitate the sowing and proper distribution of seed at appropriate depth. Pearl millet grown on deep, well-drained permeable soils usually develops extensive root systems. Mature plant roots may penetrate to a depth of 3–5 ft in an ideal soil. Root development can be severely restricted in soils having excessively high or low soil moisture levels, and hard pan and compacted layers. Soils seriously infected with witchweed or *striga* must be avoided. In semiarid and arid conditions of India, pearl millet is extensively grown in light-textured red sandy, red loamy, alluvial, and coastal alluvial soils as well as on mixed black and red and medium black soils. It is also grown on medium black soils, deep alluvial loams, and on sandy and gravelly soils of poor fertility with low organic matter content, but the yield is low.

Summer plowing is advantageous to kill the weed seeds and hibernating insects and pathogens by exposing them to the heat. Initial plowing should be carried out at optimum moisture range to get good tilth and should be avoided when moisture is in excess. Number and depth of plowings depend on weed intensity. For rainy season crop, with onset of rains in May–June, the field is plowed once or twice to obtain a good tilth.

Harrowing of soil should invariably follow after each plowing to reduce the clod size. After the initial plowing, the subsequent plowing and harrowing are carried out when the moisture content of the clods is reduced. The number of plowing should be minimized to reduce the cost of cultivation. Tillage operations should be repeated when the fields are heavily infested with perennial weeds.

3.11 VARIETIES

Pearl millet varieties recommended for fodder production in different states are given in Table 3.1.

TABLE 3.1 Pearl Millet Varieties Recommended for Fodder Production.

State	Varieties/hybrids
Uttar Pradesh, Punjab, Haryana, Madhya Pradesh, and Rajasthan	Pusa Moti, UPFB 1, T 55, S 530, A 1/30, Rajko, AVKB-19 Hybrids: NB 3, NB 17, NB 18, NB 21, NB 25, PHB 12, MH 30, BJ 105, Composite 6, K 674, K 677, L 72, L 74, Anand S 11, Raj Bajra Chari 2, Haryana, Composite 10, Nandi 32, MH-564, Nandi-8, PB 106, RHB 90, DRSB-2, Proagro-1, Pusa 383, Pusa 605, Pusa 415, MLHB-44, HAB-9, Pusa Composite 334
Maharashtra and Gujarat	Malbandro, G 2, G 5 (drought resistant)
Tamil Nadu	Co 1, Co 2, Co 8, Nad Kumbu, TNSC1 (Chumbu)
Karnataka	B 247
For saline soils	DL 454, DL 532, DL 36
Rainfed: entire bajra tract, Rajasthan and Gujarat	Rajko

Characteristics of some of the high-yielding popular hybrids and varieties.

3.11.1 FBC16

It is multicut, resistant to major diseases, high voluntary dry matter intake, and low concentration of oxalates. Its yield potential is 70–80 t green forage/ha. It is recommended for growing in plains of Punjab.

3.11.2 GIANT BAJRA

The variety was developed by MPKV, Rahuri by intervarietal hybridization between Australian and local bajra from Dhule district followed by selections. The variety has been recommended for cultivation in entire bajra-growing area. Plants are leafy with profuse tillering and have 9–10% protein at boot stage. The variety is good for hay and silage making. It is moderately resistant to downy mildew and ergot diseases. The green fodder yield is 50–75 t/ha.

3.11.3 RAJ BAJRA CHARI 2

The variety was developed by RAU, Jobner after two cycles of full sib selection in a population created through random mating among 20 crosses of four inbreds (originating from West Africa). It has been notified for cultivation for entire bajra-growing area. The green fodder yield is 30–45 t/ha and is resistant to foliar diseases and insect-pests. At ear-emergence stage, internodes are completely covered (enclosed) in the leaf sheath and the leaves are broad and shining.

3.11.4 CO8

The variety was bred by Tamil Nadu Agricultural University (TNAU), Coimbatore by hybridization (732 A × Sweet Giant Bajra) followed by pedigree selection. It was released for entire bajra-growing areas of the country. It is ready for fodder harvest in 50–55 days and produces green fodder to the tune of 30 t/ha. It has soft stem with high leaf stem ratio and is highly palatable. The variety has pale yellow-green bristles on panicles at flowering.

3.11.5 TNSC1

The variety was bred by TNAU, Coimbatore and recommended for cultivation in the entire bajra-growing areas of the country. The variety provides 27–40 t/ha green fodder and is resistant to foliar diseases and insect-pests.

3.11.6 APFB2

The variety was developed by recurrent selection in the randomly mated population at Acharya N. G. Ranga Agricultural University (ANGRAU), Hyderabad. It was recommended for cultivation in Andhra Pradesh. It belongs to early-maturity group, nonlodging, fertilizer responsive, best suited to summer, and early kharif sowings. The plant height is 160–180 cm providing green fodder yield 25 t/ha and dry fodder yield 5.5 t/ha in a single cut. The variety is useful for multicut also in the summer season.

3.11.7 PROAGRO NO. 1 (FMH3)

This variety was developed by Proagro Seed Company, Hyderabad through hybridization of PSP-21 × PP-23. The variety is recommended for cultivation throughout the pearl millet-growing areas of the country. The plants require 50–55 days for flowering and mature in 90–95 days. The variety is highly resistant to downy mildew and provides 75 t/ha green fodder in multicut system and 36 t/ha in single cut system.

3.11.8 GUJARAT FODDER BAJRA-1 (GFB1)

This variety has been bred by GAU, Anand.

3.11.9 FBC16

The variety has been bred by Punjab Agriculture University (PAU), Ludhiana and notified for cultivation in the entire northwest India. This is a multicut variety, resistant to major diseases. The variety has low concentration of oxalates and high voluntary dry matter intake by the animals. The green fodder yield potential is 70–80 t/ha.

3.11.10 AVIKA BAJRA CHARI (AVKB19)

The variety has been bred by IGFRI-RRS, Avikanagar by selection from material collected from Nagore, Rajasthan. The variety is recommended for cultivation in the state of western Uttar Pradesh, Rajasthan, Haryana, Punjab, and Terai region of Uttarakhand. The variety is a dual purpose with green fodder yield potential of 36.7 t/ha, dry fodder 8.8 t/ha, and 10.2 quintals/ha seed yield.

3.12 SOWING TIME

The sowing time of pearl millet is an important aspect in increasing the crop yield. The sowing time is also related to soil moisture and soil temperature, as well as the distribution of rainfall. In general, the rainy season crop should be sown immediately after commencement of monsoon and/or when soil has adequate moisture. The seeds should be sown when there is moisture in the soil up to a depth of 15–18 cm. If irrigation facility is available, irrigating the field for sowing just before the onset of rainy season, and thus advancing the sowing substantially increases pearl millet yield.

Most of pearl millet in India is grown in rainy (*kharif*) season (June–September), is also cultivated during summer (February–May) in Gujarat, Rajasthan, and Uttar Pradesh, and during post-rainy (*rabi*) season (November–February) at a small scale in Maharashtra and Gujarat. During *kharif* season, pearl millet is largely grown as rainfed crop except in some areas in eastern Rajasthan, southern Haryana, and western Uttar Pradesh where supplemental irrigation is provided in case of shortage of rainfall during the crop season. Summer season pearl millet is cultivated as an irrigated crop under high levels of agronomic management. Soil temperatures should be at least 18°C or warmer before pearl millet is sown. It germinates well at soil temperatures of 20–30°C.

Pearl millet is also transplanted in some parts of Gujarat, Tamil Nadu, Andhra Pradesh, and Karnataka. This practice is common under irrigated conditions. Pearl millet nurseries are raised in well-fertilized raised seedbeds. Healthy seedlings (15–20-days old) are transplanted in the water-soaked fields. A nursery of 0.03 ha is sufficient to supply seedlings to plant 1 ha main field.

3.13 SEED TREATMENT

The seed should be treated with appropriate chemicals prior to planting to prevent seed-borne diseases as well as soil pests, which are common in the field. The seeds are also treated with some biofertilizers for easy and enhanced availability of important nutrients such as nitrogen and phosphorus. Seed hardening is practiced for better germination. Pearl millet seeds should be treated with recommended agrochemicals following the given directions for use. This seed treatment prior to sowing will help in preventing soil-inhabiting insects and soil-borne diseases. Seed treatment is particularly important where dry sowing prior to rains is practiced. The following are the seed treatment practices (Khairwal et al., 2007):

- Soak the seeds in 2% (20 g in 1 L of water) potassium dihydrogen phosphate solution for 6 h. Use 350 mL of solution for soaking 1 kg of seed. Dry the seed in shade to original moisture level.
- Soaking seeds in 1% calcium chloride or in plain water for 6 h before sowing results in increased yield under drought conditions.
- Seed treatment with biopesticides such as *Trichoderma harzianum* or *Trichoderma viride* using 4 g/kg or with Thiram 75% WP at 0.75 g/kg seed or with Thiram 75% dust/Captan 75% at 3 g/kg of seed will help against soil-borne diseases.
- The seeds of pearl millet are treated with 300 mesh sulfur powder at 4 g of sulfur per kilogram of seeds for controlling the smut disease.
- Soak seeds in 10% salt solution (1 kg of common salt in 10 L of water). Remove ergot-affected seeds which float to reduce the incidence of ergot disease.
- Seed treatment with metalaxyl (Apron 35 SD) at 6 g/kg seed controls downy mildew.
- Treating the seeds with three packets each of *Azospirillum* (600 g) and *Phosphobacterium* will enhance the availability of nitrogen and phosphorus.
- Soaking the seed for 1 h in 1% of 2-chloroethanol plus 0.5% sodium hypochlorite solution is effective in increasing germination rate.
- Seed treatment is done manually for small amount of seeds. Large amount of seeds can be treated using a seed-treating drum.

3.14 SEED RATE AND SOWING METHOD

The amount of seed required per unit area depends on: optimum plant stand required per unit area, plant type, that is, tall or short plants, seed mass, and quality of seed in terms of germination. Seed rate should be determined based on the optimum number of plants required per unit area (acre or ha or sq ft or sq m) for good yields.

Use certified seed for sowing: for getting higher yields, it is necessary to use new hybrid seed every year. Before sowing, seed lot must be tested for its germination percentage. A seed rate of 8–10 kg/ha is sufficient for fodder production of pearl millet sown by drilling in 30 cm rows. The crop is sown by broadcast with 10–15 kg seed/ha. Seeds should be sown about 2–3 cm deep. The seeds are sown either by broadcasting seed manually or sowing behind country plow using *pora* (country seed drill), or an improved seed drill which may or may not be fitted with hoppers for fertilizer application, or mechanical seed drills attached to a tractor. Manual sowing by broadcasting the seed is practiced when the area to be sown is small. After broadcasting, the seeds are covered by running a brush harrow. The germination may not be uniform in hand sowing, so one has to use a higher seed rate to get the optimum plant stand in the field.

A country seed drill called *pora*, that is, a wooden hopper attached to a hollow bamboo is used to sow the seeds in the shallow furrows opened by the tines of a country plow. A larger area can be covered with this method of sowing. The best method is to sow the hardened seeds of pearl millet at 5 cm depth with seed-cum-fertilizer drill to ensure uniform depth of sowing and fertilizer application before the onset of monsoon.

Bajra is generally sown behind plow or by broadcast method. These methods are quite unsatisfactory and generally lead to poor germination and consequently poor yield. Sowing bajra with seed drill is the best method. It not only ensures best germination but also uniform plant population as well. About 2 kg seeds of pearl millet are sown in 500–600 m^2 area in nursery to get seedlings for 1 ha. The seeds are sown in flat beds (1.20 m × 7.50 m) in row 10 cm apart and at 1.5 cm depth. To give better start to the seedlings, apply 25–30 kg calcium ammonium nitrate in the nursery. The seedlings are uprooted and transplanted after 3 weeks. While uprooting the seedlings, keep the nursery wet just to avoid root injury. Remove the top portion above the growing point so as to minimize the transpiration from the seedlings. Transplanting should be done preferably on rainy days. If it is not raining, irrigate the field to help the seedlings to establish themselves. Transplant one seedling per hole in rows keeping 50-cm space between rows and 10-cm space between plants. Transplanting from third week of July to second week of August gives good results.

3.15 INTERCROPPING

Intercropping refers to growing more than one crop in the same land area in rows of definite proportion and pattern. Pearl millet and groundnut intercropping system is recommended to farmers to meet the fodder needs of cattle and milch animals. With particular reference to dry land agriculture, an intercropping system needs to be designed in such a way that in the case of unfavorable weather, at least one crop will survive to give economic yields.

Thus, intercropping system should provide for the necessary insurance against unpredictable weather. In case the year happens to be normal with respect to rainfall, the intercropping system, as a whole, should prove to be more profitable than growing either of the crops alone.

An ideal intercropping system should aim to:

- Produce higher yields per unit area through better use of natural resources.
- Offer greater stability in production under adverse weather conditions and disease and insect infestation.
- Meet the domestic needs of the farmer.
- Provide an equitable distribution of farm resources.

While maintaining the yield levels of the sole crop of pearl millet, additional yields with the intercropping component have been realized under various systems. Since a food legume is involved in most of the systems, it will not only enhance the income of the farmer but also provide the much-needed protein to supplement the predominantly cereal diet of farmers and help improve soil fertility. The suggested state-wise pearl millet-based intercropping for India is given in Table 3.2.

3.16 CROP MIXTURE

Mixed cropping refers to simultaneously growing more than one crop in the same land area as a mixture. Unlike in intercropping system, in mixed cropping the crops are grown without any definite pattern. Mixed cropping of pearl millet–pigeon pea or cluster bean is most common. Mixtures with green gram, black gram, cowpea, and even with sorghum and other cereals, vegetables, etc. during rainy season are practiced under different

TABLE 3.2 Suggested Intercropping Practices in Pearl Millet.

State	Suggested intercropping	State	Suggested intercropping
Eastern Rajasthan	Pearl millet + cluster bean	Maharashtra	Pearl millet + pigeon pea/soybean
	Pearl millet + cowpea		Pearl millet + black gram
	Pearl millet + green gram		Pearl millet + green gram
Western Rajasthan	Pearl millet + moth bean		Pearl millet + cowpea
	Pearl millet + cluster bean		Pearl millet + moth bean
Haryana	Pearl millet + cluster bean		Pearl millet + sunflower
	Pearl millet + green gram	Karnataka	Pearl millet + pigeon pea/soybean
	Pearl millet + cowpea		Pearl millet + green gram
Uttar Pradesh	Pearl millet + green gram		Pearl millet + sunflower
	Pearl millet + cowpea	Andhra Pradesh	Pearl millet + pigeon pea/soybean
	Pearl millet + sesame		Pearl millet + green gram
Madhya Pradesh	Pearl millet + pigeon pea		Pearl millet + sunflower
	Pearl millet + cowpea		Pearl millet + groundnut
	Pearl millet + soybean	Tamil Nadu	Pearl millet + pigeon pea/soybean
Gujarat	Pearl millet + green gram		Pearl millet + green gram
	Pearl millet + cowpea		Pearl millet + cowpea
	Pearl millet + sesame		Pearl millet + sunflower

situations. Mixed cropping is practiced in traditional subsistence farming to meet the domestic needs of the farmer's family. Thus, the number of crops mixed varies depending on the family needs. Even though crops in the mixed cropping meet the farmer's family needs, the yield of crops will be low due to the competition between the crops for water, light, nutrients, etc.

3.17 CROP SEQUENCE

Pearl millet crop is mostly grown as a rainfed monsoon crop during rainy season (June–July to October–November) and also, to a limited extent, as an irrigated summer season crop (January–May) in India. Pearl millet is often grown in rotation with sorghum, groundnut, cotton, foxtail millet, finger millet, castor, and sometimes, in southern India, with rice. If the pearl millet crop is sown early in May, it can be followed in the same year by horse gram. In areas where cotton and sorghum are grown, the rotations followed may be pearl millet–cotton–sorghum or pearl millet–sorghum–cotton (a 3-year rotation). In sandy soils, pearl millet is normally grown continuously year after year, which deteriorates soil health.

Cluster bean (*guar*)–pearl millet crop sequence with crop residue incorporation has significantly increased the productivity in arid zone of western Rajasthan where fallow–pearl millet/pearl millet after pearl millet crop sequence are practiced.

In Punjab, the dry land rotation may be small grain–millet–fallow. In irrigated lands, pearl millet is rotated with chickpea, fodder sorghum, and wheat. In dry and light soils of Rajasthan, southern Punjab, Haryana, and northern Gujarat, pearl millet is often rotated with a pulse such as moth bean or green gram, or is followed by fallow, sesame, potato, mustard, and *guar*. Sesame crop may be low yielding and may be replaced by castor or groundnut.

In some regions of Tamil Nadu, the rotation may be more complex: pearl millet–finger millet–groundnut–rice–sugarcane in a 3-year rotation, with irrigation. In red soils of Karnataka, pearl millet and finger millet rotation is practiced, though pearl millet might not be grown every year. In coarse gravelly soils, castor may follow pearl millet. Rotation of cultivars also should be adopted to avoid downy mildew disease problem. Pearl millet hybrids and open pollinated varieties should be used in alternate years/seasons. It is advised not to grow the same hybrid or open pollinated variety continuously on the same piece of land.

3.18 MANURES AND FERTILIZERS APPLICATION

The fertilizer requirement of local varieties of pearl millet can easily be met by the application of 10–15 t of compost or farm yard manure per hectare. But the nutrient supply for the high yielding varieties and hybrids should be supplemented with inorganic fertilizers. Amount of fertilizer should be given on the basis of soil test value for maximum profit. To get good fodder crop, 40–60 kg N/ha and 20–30 kg P_2O_5/ha may also be applied at the time of sowing. In soils deficient in K, 30–40 kg K_2O/ha should be applied. In rainfed conditions, foliar spray of 3% urea is also recommended. Azospirillum, a biofertilizer, is recommended at the rate of 2 kg/ha mixed with each of 25 kg farm yard manure and soil, and applied at the final plowing or at sowing. This will not only help to reduce the rate of nitrogen fertilizer application, but also enhance better utilization of applied nitrogen by the plants.

3.19 WATER MANAGEMENT

Bajra is mainly grown under rainfed situation. Only about 8% of bajra area is irrigated in India. The crop is exposed to drought conditions very often during its growth and consequently the yields are lower than the potential yield. So, if the farmers want higher bajra yields even during the rainy season, the crop should be irrigated, if water is available and if there is any dry spell. Irrigation helps bajra crop to make efficient use of inputs, increases yield, improves quality of grain, improves reliability and reduces risk, and increases profitability. Adequate soil moisture at sowing helps assure uniform plant stand and contributes to early plant growth. Preplanting irrigation can supply this moisture when early rains do not reach the root zone prior to planting. Allowing the seed to remain in dry soil for several days may result in poor germination and seedling vigor. So, irrigation prior to sowing is recommended if water is available. Whether irrigating before or after planting, apply no more water than required to reach the effective root zone. Encourage deep root system by maintaining only moderate soil moisture levels during early vegetative growth.

Moderate plant moisture stress during early vegetative growth normally does not reduce grain yield significantly. There is little quantified information about pearl millet response to irrigation during its growth. It appears that pearl millet responds less to irrigation than other grain crops. Greatest water use occurs during the flowering and soft dough stages. Irrigation intervals and the amount of water to be applied is determined by rainfall, soil water holding characteristics, plant rooting depth, and other climatic conditions such as air temperature, etc. The visible signs for irrigating the crop or moisture requirement of the crop are drooping and withering of leaves in the morning, and cracking of the soil surface of the field. Plan the irrigation to provide adequate soil moisture during these periods of high water requirement.

3.20 WEED MANAGEMENT

Pearl millet being mainly grown during the rainy season encounters several weeds which grow luxuriantly and dominate during this season more as compared to *rabi*/summer season (Table 3.3). Weeds in pearl millet crop compete with the crop for nutrients, water, and light as with other crops. Weeds are more competitive when moisture is limiting and especially under drought situation when young, grain pearl millet does not compete well with weeds.

The damage is severe during 3–5 weeks after sowing. Therefore, timely control of weeds is essential to get higher yields. Interculture the crop 3–5 weeks after sowing. Avoid deep hoeing near the plants, so that their roots are not damaged. A wheel hoe, triphali, or hand hoe can be used for interculture. Under such circumstances the only effective way to control weeds is the use of herbicides. Preemergence application of atrazine at the rate of 0.5 kg active ingredient per hectare in 800 L of water controls most of the monocot and dicot weeds.

Several weeds are associated with pearl millet crop. The type of weeds varies from place to place depending on soil type, season, and environment where pearl millet is grown. These weeds comprise diverse plant species of annual grasses, seasonal broad-leaved weeds, and sedges. Weed problems in pearl millet include perennial grasses such as Johnson grass and Bermuda grass, annual grasses such as crabgrass and goose grass, and many broad-leaved weeds (Table 3.3).

TABLE 3.3 Some Common Weeds Encountered in Pearl Millet Fields.

Common name	Botanical name	Family
Broadleaf		
Cock's comb	*Celosia argentea* L.	Amaranthaceae
Common purslane	*Portulaca oleracea* L.	Portulacaceae
False amaranth	*Digera arvensis* Forssk	Amaranthaceae
Horse purslane	*Trianthema portulacastrum* L.	Aizoaceae
Pigweed	*Amaranthus viridis* (Hook. F.)	Amaranthaceae
Swine cress	*Coronopus didymus* L. Sm.	Brassicaceae
Black nightshade	*Solanum nigrum* L.	Solanaceae
Grasses		
Rabbit/crow foot grass	*Dactyloctenium aegyptium* Willd	Poaceae
Goose grass	*Eleusine indica* L. Gaertner	Poaceae
Barnyard grass	*Echinochloa crusgalli* L. Scop	Poaceae
Crabgrass	*Digiteria sanguinalis* L. Scop	Poaceae
Bermuda grass	*Cynodon dactylon*	Poaceae
Sedges		
Purple nutsedge	*Cyperus rotundus* L.	Cyperaceae
Flat sedge	*Cyperus iria* L.	Cyperaceae
Parasitic		
Witchweed	*Striga asiatica* (L.) Kuntze	Orobanchaceae

3.21 DISEASE MANAGEMENT

Among the various factors responsible for low production of bajra diseases are major limiting factors. Diseases that occur on pearl millet crop result in considerable economic losses because most of these diseases affect pearl millet panicles and the grains. These economically important diseases are downy mildew, ergot, smut, and rust. Symptoms of important diseases and their suitable management are given below.

3.21.1 DOWNY MILDEW

Downy mildew of pearl millet, caused by *Sclerospora graminicola*, sometimes referred to as green ear, is the most destructive disease of pearl millet. The symptoms of disease may be seen at an early stage on leaves. Infected leaves become yellow or white and downy with yellowish longitudinal streaks extending from the tip to the base. Such leaves soon turn brown and tear off at the streaks. The fungal growth can be seen on the lower surface of leaves in humid weather. The most distinguishing symptom of disease is transformation of ears into leafy whorl-type structures. The name green ear disease is derived from this stage of the disease.

3.21.1.1 MANAGEMENT

Management practices should aim at reducing the movement of primary soil and seed-borne fungal inoculums, and the secondary spread of the fungus within and between pearl millet fields. This can be achieved by the combination of the three disease management practices—cultural, chemical, and host plant resistance.

- Use tolerant/resistant varieties/hybrids, for example, IPFM-5, CO-8, Giant Bajra, IP-36B, IP-8556, IP-11890, L-72, MP-260, MP-261, PCB-150, PCB-151, REB-1, TNSC-4, TNSC-5, TNSC-6, UUJ-2, NBH 5, PGB 10 PHB 14 etc.
- Use of disease-free seed and effective removal of infested plant material from the field after the harvest of the crop is essential to reduce the primary inoculums in the soil. Downy mildew-infected plant material should be burned, or if feasible, the field should be plowed deeply to bury the plant material.
- Seed treatment with fungicide such as agrosan GN, Thiram (2.5 g fungicide/kg of seed).
- Foliar spray of dithane Z 78 @ 0.2% or copper oxychloride @ 0.35% at boot-leaf stage is effective in its control.

3.21.2 ERGOT

This disease is caused by a fungus known as *Claviceps microcephala*. The disease first appears on the ears in the form of honey-like pinkish liquid which is full of fungus spores, causing spread of the disease. The liquid turns brown and sticky. In the later stages, fungus sclerotia (ergots) appear as brown to black and elongate structures. They possess toxic alkaloids harmful to human and cattle health. These sclerotia fall down in the field and remain in soil, causing infection in subsequent crops. It is important in fodder bajra seed production.

3.21.2.1 MANAGEMENT

- The preventive measures include a plowing during summer, avoiding late planting, use certified seeds, and their treatment with 20% common salt solution followed by washing with fresh water and then treating with agrosan GN, Thiram, or ceresan @ 3 g/kg of seed.
- Spraying the crop with 0.15% Thiram at boot-leaf stage is effective in its management.
- Use of resistant cultivars is the most cost-effective method for the control of ergot disease.
- Eradicate the weeds such as *Cenchrus ciliaris* and *Panicum antidotale* from around pearl millet fields.

3.21.3 SMUT

Smut of pearl millet, caused by *Tolyposporium pesiscillaria*, is one of the important diseases affecting ears. It is common under humid conditions in all the states where pearl millet is grown. The diseased kernels are green in the beginning and generally larger in size than the healthy ones. These are full of fungal spores.

3.21.3.1 MANAGEMENT

- Deep plowing during summer and following a 3-year crop rotation. Use certified seeds, and treat them with fungicides such as agrosan GN or Thiram or cerasan @ 3 g/kg of seed.
- Spraying the crop with 0.15% vitavex (1.5 kg vitavex mixed in 1000 L of water/ha) at boot-leaf stage followed by 1–2 sprays at 15 days intervals is recommended.
- Intercropping of mung bean with pearl millet reduces the smut disease.

3.21.4 RUST

This disease is caused by *Puccinia penniseti* has been observed throughout India. The disease is seen on both surfaces of leaves. The rust pustules are small, orange, and full of spores. On touching these pustules, one can see orange-colored spores sticking on the finger tips. In later stages, black-colored pustules can be seen. In severe infection, leaves are completely dried.

3.21.4.1 MANAGEMENT

- Rust-causing fungi develop on many grass weeds. These weeds help in spreading the disease to pearl millet crop. So, good weed control practice helps reduce rust in pearl millet.
- Spray the crop with 0.2% zineb or Mancozeb 75 WP three to four times to provide good protection.
- Resistant and moderately resistant to rust varieties are APFM-2, APFM-5, APS-1, CO-8, and Giant Bajra.

3.21.5 LEAF BLAST

This disease is caused by *Pyricularia setariae.* The symptoms of this disease can be seen on the lower surface in the form of light to dark brown, boat-shaped lesion. Sometimes several spots coalesce and form larger patches. In severe infection, leaves are completely dried.

3.21.5.1 MANAGEMENT

The disease can be controlled effectively by two or three sprays of 0.2% zineb.

3.22 INSECT-PEST MANAGEMENT

Pearl millet crop is subject to attack by a number of pests. A brief description of the major ones, their nature of damage, and suitable control measures are given below.

3.22.1 WHITE GRUBS

Holotrichia sp., *Anomala* sp. (Melolonthidae: Coleoptera).

3.22.1.1 DAMAGE SYMPTOMS

Holotrichia consanguinea devastates pearl millet crop in large areas in central India. The grubs cut the roots resulting in wilting of plants in patches and die. Even three to four grubs may attack the same plant. Infested seedlings remain stunted and produce no seeds.

3.22.1.2 MANAGEMENT

- Intercropping of pearl millet with pulses such as green gram, cluster been, and cowpea significantly reduces the white grub damage on pearl millet crop.
- Soil drenching of imidacloprid 17.8% SL @ 300 mL/ha or mixing chlorpyriphos 20% emulsifiable concentrate (EC) or quinalphos 25% EC @ 4 l/ha along with irrigation water in standing crop after 3 weeks of emergence of beetle is suggested.

3.22.2 SHOOT FLY

Atherigona approximata (Muscidae: Diptera).

3.22.2.1 DAMAGE SYMPTOMS

Shoot fly is a serious pest of pearl millet in north India. Plants up to 3 weeks of age are favorite of the pest. The fly lays eggs on the lower surface of the leaves or near the base of the plants. Within 2 days or so, the eggs hatch and the tiny maggots creep down under the leaf sheaths and cut the central growing point of the plant which results in the dead heart.

3.22.2.2 MANAGEMENT

- Removal of the insect-infested seedlings with dead hearts and keeps the optimum plant stand in the field.
- Plow the field after harvest to remove and destroy the stubbles.
- Setup the low cost fish meal traps @ 4/acre till the crop is 30 days old.
- Seed treatment with imidacloprid 48% FS 12 mL/kg seed or imidacloprid 70% WS 10 g/kg seed.
- At the time of planting, 20 kg/ha of carbofuran 3G granules may be applied in furrows before sowing to protect seedlings against shoot fly. When the shoot fly damage reaches 5–10% of plants with dead hearts, the crop may be sprayed with cypermethrin 10% EC @ 1 mL/L.

3.22.3 STEM BORER

Chilo partellus (Crambidae: Lepidoptera).

3.22.3.1 DAMAGE SYMPTOMS

It infests the crop a month after sowing and up to emergence of earhead. Central shoot withering leading to dead heart is the typical damage symptom. Bore holes visible on the stem near the nodes. Young larva crawls and feeds on tender folded leaves causing typical shot hole symptom. Parts of stem may show internally tunneling caterpillars.

3.22.3.2 MANAGEMENT

- The stubbles should be plowed up during winter and burnt to destroy the hibernating larvae.
- Dead hearts should be pulled out and used as fodder or buried in manure pits.
- Sow lablab or dolichos as an intercrop in the ratio of 4:1 to minimize the stem borer damage.
- Setup light trap till midnight to attract and kill the stem borer moths.

- Biocontrol agents, namely, *Trichogramma chilonis* (egg parasitoids) *minutum*, *Bracon chinensis*, and *Apanteles flavipes* (larval parasitoids), should be encouraged.
- Carbofuran 3G granules may be applied in the whorls @ 8–12 kg a.i./ha or the entire field can be sprayed with metasystox @ 2 mL/L.

3.22.4 GRAIN MIDGE

Geromyia penniseti (Cecidomyiidae: Diptera).

3.22.4.1 DAMAGE SYMPTOMS

The maggot feeds on the developing grains and pupates there. Maggots destroy the ovaries seriously, affecting the development of seeds, leading to chaffy panicles. White pupal cases protruding out from the chaffy grains with exit holes are seen. Pupal cases can be seen attached to the glumes of damaged spikelet. Midge complete four to five generations in a season with overlapping generations.

3.22.4.2 MANAGEMENT

The crop may be sprayed at the 50% flowering stage (one midge/panicle) with cypermethrin 25 EC @ 0.5 mL/L. Carbaryl 10% dust @ 20–25 kg/ha at 50% flowering and grain formation stage is also suggested in midge endemic areas.

3.22.5 STINK BUG

Nezara viridula (Pentatomidae: Hemiptera).

3.22.5.1 DAMAGE SYMPTOMS

Grains become chaffy or spotted black and get shriveled. A stinking smell emanates from the bug.

3.22.5.2 MANAGEMENT

Apply any one of the insecticides at 25 kg/ha carbaryl 10D, malathion 5D, or spray carbaryl 50 WP 750 g/ha at 50% flowering stage.

3.22.6 EARHEAD BUG

Calocoris angustatus (Miridae: Hemiptera).

3.22.6.1 DAMAGE SYMPTOMS

The adults and nymphs damage the earheads by feeding on them. They suck the juice from the grains when they are in the milky stage. The sucked out grains, shrink and turn black in color and become ill filled (or) chaffy. Older grain shows distinct feeding punctures that reduce grain quality.

3.22.6.2 MANAGEMENT

- Spray NSKE 5%.
- Spray azadirachtin 1%.
- Dust with carbaryl 10% at 12 kg/ha (or) quinalphos 1.5% 12 kg/ha synchronizing during milky stage.

3.22.7 RED HAIRY CATERPILLAR

Amsacta albistriga, A. moorei (Arctiidae: Lepidoptera).

3.22.7.1 DAMAGE SYMPTOMS

The larvae feed on the leaves gregariously by scrapping the under surface of tender leaflets leaving the upper epidermal layer intact in early stages. Later, they feed voraciously on the leaves and main stem of plants. They march from field to field gregariously. Severely affected field looks as if grazed by cattle.

3.22.7.2 MANAGEMENT

- Use light traps.
- Irrigate once to avoid prolonged midseason drought to prevent preharvest infestation.
- Dig the trenches of 1-in. depth between the fields and dust the trenches to kill the larvae in pits.
- Spray *Bacillus thuringiensis* @ 400 g/acre.

3.22.8 TERMITES

Odontotermes spp., *Microtermes* spp., *Macrotermes* spp. (Termitidae: Isoptera).

3.22.8.1 DAMAGE SYMPTOMS

Termites also attack the roots and damaged plants topple. They eventually disrupt the movement of nutrients and water through the vascular system resulting in death of the plant. The dead plants are sheathed with soil. The pest is more serious in dry areas.

3.22.8.2 MANAGEMENT

For areas of regular termite occurrence, the soil should be mixed with chlorpyriphos 5% dust @ 35 kg/ha at the time of sowing. When the incidence of pest is noticed in standing crop, mix chlorpyriphos 20% EC @ 4 L/ha with 50 kg of soil and broadcast it evenly in 1 ha followed by light irrigation.

3.23 HARVESTING

The crop is ready for harvest in 60–70 days after sowing. Harvesting at 50% flowering stage is ideal. In multicut varieties, first cut is taken 50–55 days after sowing (a little earlier to 50% flowering) and subsequent cuts at 35–40 days interval.

The ideal moisture content for harvesting grain pearl millet is below 20%. Thin stems, heavy panicles, and profuse tillering, which are inherent traits of varieties, may result in lodging of the plants. The seeds in the panicles of lodged plants germinate in the moist field, and thus affect grain yield and quality. Hence, the plants are tied together to prevent lodging. Most standability concerns develop when growers leave pearl millet in the field until grain moisture is below 14%.

When the crop matures, the leaves turn yellowish and present a nearly dried up appearance. The grains are hard and firm. The usual practice of harvesting pearl millet is cutting the earheads first and the stalks later. The stalks (straw) are cut after a week, allowed to dry and then stacked. The harvested earheads are dried before threshing. The grain is separated from the earheads by using a mechanical thresher or by drawing a stone roller over the earheads or trampling with cattle. The separated grain is then cleaned by winnowing and dried. Conventional grain dryers or use of natural air drying are options for drying the grain. Grain at or below 14% moisture is considered dry. For long-term storage (more than 6 months), grain moisture content should be less than 12% after which the grains may be bagged and stored in a moisture-proof store.

3.24 YIELD

The yield of bajra with improved cultural practices is nearly 30–35 quintals of grain and about 100 quintals of dry stover from a hectare of crop under irrigated conditions and about 12–15 quintals of grain and 70–75 quintals of dry stover from a rainfed crop. The green fodder yield varies from 35–40 t/ha.

3.25 SEED PRODUCTION

Bajra is a highly cross-pollinated crop with 80% of cross-pollination. The crop should be raised in isolation and seeds should be allowed to set by open pollination. The isolation distance maintained between the varieties is 400 m for foundation seed and 200 m for certified seed production.

Seed production stages:

Breeder seed→foundation seed→fertified seed.

3.26 NUTRITIVE VALUES

Pearl millet is regarded as a high-quality feed for grazing, green chop, hay, or silage. It is high in protein and energy and low in fiber and lignin concentration. Crude protein can range from 9% to 11% in unfertilized soils to 14% to 15% under nitrogen-fertilized conditions.

It is also high in calcium and iron and has balanced amino acids, but sulfur-containing amino acid concentration is low. The forage is readily consumed by livestock when used at vegetative stages. In situations where the grain is harvested, the nutritive value of the material remaining on the field drops substantially due to low leaf-to-stem ratio of the plants and standing stems that are substantially low in digestibility and nutrient concentration.

3.27 UTILIZATION

The utilization of bajra as a food crop is limited to the developing countries in Asia and Africa. It is estimated that over 93% of bajra grain is used as food, the remainder being divided between animal and poultry feed (7%). Other uses include bakery products and snacks, to a very limited extent. With a texture much like brown rice, bajra can be cooked like rice. The cooked bajra will be fluffy and delicate cereal. Bajra is traditionally used for food products such as *roti* (flat bread), *bhakri* (stiff *roti*), and porridge or gruel. Bajra flour mixed with wheat flour is used for making baking products such as breads, cakes, muffins, cookies, and biscuits.

The crop residue (stover), after the harvest of grains, is used as a dry fodder, particularly during the winter months when fodder is usually scarce. The crop residue (stover) after grain harvest is a valuable source of fodder for livestock. Utilization of grain as feed for milch animals or poultry is not significant (at present 7%) in India but in future, it will be a crop of choice for animal and poultry feed. In addition, this crop is grown especially for production of green fodder, particularly in irrigated areas near towns. Nowadays, several varieties and a few hybrids are available exclusively for forage purposes. Forage bajra is fed to animals as a green chop or hay (quickly dried fodder).

3.28 SPECIAL FEATURES (TOXICITIES)

Bajra has a high potential for accumulating toxic levels of nitrate, especially on the lower 6 in. (15 cm) of the stalks (Strickland et al., 2007). It is best to avoid grazing younger plants and to avoid overgrazing. Droughty or cold weather can stress plants and increase nitrate levels. Bajra may contain higher levels of nitrate than sorghum–Sudan grass after hot weather, however, nitrate returns to safe levels 7–14 days after a drought-ending rain (Strickland et al., 2007). Bajra feed can be diluted by mixing with low-nitrate feeds. Newman et al. (2010) observed that haying material does not reduce nitrate concentrations, but ensiling the forage can decrease nitrate levels 40–60%.

FIGURE 3.1 (See color insert.) Pearl millet.

KEYWORDS

- **pearl millet**
- **cultivation practices**
- **nutritive value**
- **utilization**
- **toxicities**

REFERENCES

Andrews, D. J.; Kumar, K. A. Pearl Millet for Food, Feed and Forage. In *Advances in agronomy*; Sparks, D. L., Ed.; Academic Press, 1992, Vol. 48, pp 90–139.

Hassanat, F. Evaluation of Pearl Millet Forage. Ph.D. Dissertation, Department of Animal Science, Mc Gill University, Montréal, Québec, Canada, 2007.

Khairwal, I. S.; Rai, K. N.; Diwakar, B.; Sharma, Y. K.; Rajpurohit, B. S.; Bindu, N.; Bhattacharjee, R. *Pearl Millet: Crop Management and Seed Production Manual*; International Crops Research Institute for the Semi-Arid Tropics: Patancheru, Andhra Pradesh, India, 2007; p 104.

Newman, Y.; Jennings, E.; Vendramini, J.; Blount, A. *Pearl Millet* (*Pennisetum glaucum*)*: Overview and Management*; IFAS Extension Publication: University of FL, 2010; p 337. http://edis.ifas.ufl.edu/ag347 (accessed Jul 31, 2014).

Strickland, G.; Selk, G.; Zhang, H. *Nitrate Toxicity in Livestock*; Oklahoma State University Extension Publication: Oklahoma State University, 2007; p 2903.

Tabosa, J. N.; Andrews, D. J.; Tavares-Fillio J. J.; Azevedo-Neto, A. D. Comparison Among Forage Millet and Sorghum Varieties in Semi-Arid Pernambuco. Brazil: Yield and Quality. *Int. Sorghum Millet Newsl.* **1999,** *40*, 3–6.

Timbo, A. L.; Davide, L. C.; Pinto, J. E. B. P.; Pireira, A. V. Protoplast Production from Napier Grass and Pearl Millet Triploid Hybrids. *Ciênc. Agrotec.* **2010,** *34* (5), 1219–1222.

USDA. GRIN-Germplasm Resources Information Network, National Germplasm Resources Laboratory: Beltsville, Maryland, 2009.

CHAPTER 4

SORGHUM (JOWER)

JANCY GUPTA* and MINU SINGH

Dairy Extension Division, NDRI, Karnal 132001, Haryana, India

**Corresponding author. E-mail: jancygupta@gmail.com*

ABSTRACT

Sorghum is a rich source of food as well as fodder in many tropical and subtropical countries. It is drought resistant and suited for areas where moisture is a limiting factor for crop growth. Sorghum is quite soft, palatable, and fast growing annual fodder crop adapted to areas up to 1500 m altitude. However, it remains green and palatable over a longer period than maize and pearl millet fodders. Sorghum fodder crops yield about 500–600 quintals of green fodder per hectare depending upon the season. The crop yields 200–300 quintals per hectare green fodder under dry land areas. On an average the seed yield of 10–12 quintals/ha can be obtained from the good crop. The crude protein content of sorghum is 7.75%; it may vary on the basis of stage of cutting and fertilizers applied. HCN is present in early stages up to 40–50 days, so harvesting should be done at after 55–60 days of sowing and irrigation facilities should be taken care of, as in drought conditions the fodder crop contains more of HCN.

4.1 BOTANICAL CLASSIFICATION

Kingdom: Plantae
Order: Poales
Family: Poaceae
Subfamily: Panicoideae
Genus: *Sorghum*
Species: *bicolor* (L.) Conrad Moench

4.2 COMMON NAME

It is also known as sorghum, chari, jowar, cholam, jonnalu, jola, or jwari.

4.3 BOTANICAL NAME

Sorghum bicolor (L.) Conrad Moench

4.4 INTRODUCTION

Sorghum is fourth in importance among the world's leading cereals. Among the sorghum growing countries India ranks first in acreage but second in production, and the United States is the largest producer of sorghum in world. Sorghum occupies the first position among forage crops and in addition supplies significantly large quantity of stover from grain crop for livestock (Bimbraw, 2013).

Sorghum is a rich source of food as well as fodder in many tropical/ subtropical countries. It is drought resistant and suited for areas where moisture is a limiting factor for crop growth. The great advantage of sorghum is that it can become dormant under adverse conditions and resume growth after relatively severe drought (Bimbraw, 2013). Sorghum is quite soft, palatable, and fast-growing annual fodder crop adapted to areas up to 1500 m altitude. However, it remains green and palatable over a longer period than maize and pearl millet fodders (Bimbraw, 2013).

4.5 PLANT DESCRIPTION

Sorghum is planted during Kharif season, in summer under irrigated conditions. Sorghum is available in single cut and multicut verities. Single cut is preferred over multicut in Kharif season, as it fits well in sorghum–wheat crop sequence. Multicut sorghum is popular for intensive fodder production. Sorghum is suitable for silage, hay making, and as green fodder (Panda, 2014).

Its ability of fast growth, high biomass accumulation, and dry matter content and wide adaptability beside drought withstanding ability, favors its cultivation as fodder crop.

4.6 ORIGIN AND DISTRIBUTION

Sorghum is said to be originated in East Central Africa, in or near Ethiopia or Sudan. It is also cultivated in other countries such as Southern Europe, Russia, Near East, Far East, Australia, China, Pakistan, warmer parts of United States and Latin America. The United States is the largest producer of sorghum in the world followed by India. In India, sorghum was introduced since 1300 A.D. In India, it is cultivated in most parts of North India, Andhra Pradesh, Tamil Nadu, Karnataka, Maharashtra, Madhya Pradesh, and parts of Gujarat (Bimbraw, 2013).

4.7 PLANT CHARACTERISTICS

Sorghum can be grown in arid regions and tolerate high temperature and light intensities. Sorghum can synthesize heat-shock proteins rapidly when temperature rises, so it can tolerate heat stress. These heat-shock proteins can prevent enzymes to be denatured and make them more thermostable. Sorghum plants carry out C4 photosynthesis and can tolerate high intensity. When carbon dioxide is absorbed in the leaf cells, then it combines with molecules in the cells to form a molecule of four carbon atoms (Bimbraw, 2013).

Sorghum is an annual or short-lived perennial 0.5–6 m in height. The base diameter of the stem varies from 0.5 to 5 cm. Leaves are 30–100-cm long and up to 12-cm wide and are generally waxy. Roots are more linear and fibrous. Roots are highly efficient to exhaust most of the nutrients and moisture available in the soil. Stem of sorghum may be juicy and sweet (Bimbraw, 2013).

4.8 CLIMATIC REQUIREMENT

Optimum climate requirement for sorghum cultivation is hot and dry climate. It grows well in warm condition (26–30°C) with annual rainfall 60–100 cm. For proper germination of seed 6–10°C is the suitable temperature.

4.9 SOIL AND ITS PREPARATION

Sorghum can be cultivated in all types of soils varying from sandy to clay loam under irrigated conditions; however it thrives well on heavy soils.

The field should be well drained. Alkaline soils are not suitable for cultivation. As the water requirement is very low, so it can be cultivated in areas where irrigation facilities are scarce. It can also tolerate excess moisture; therefore can be grown in flood prone areas. Only care should be taken that water logging does not continue for longer period (Bimbraw, 2013).

Normally for preparing land for plantation two to four operations by desi plow or blade harrow is needed for harrowing. And by tractor two to four harrowing are sufficient.

4.10 VARIETIES

Some of the common varieties prevalent in India are PC-1, PC-6, PC-9, PC-23, HC-136, HC-171, HC-260, SL-44, MP-Chari, UP-Chari, Raj Chari, PSC-1, Pant Chari-5, CO-27, and Meethi Sudan (Bimbraw, 2013; Anonymous, 2012; Anonymous, 2009).

4.11 SOWING TIME

Sorghum can be planted from the last week of March to mid July under irrigated conditions. Generally, crop is sown during second fortnight of June to mid of July (Bimbraw, 2013).

4.12 SEED TREATMENT AND SEED INOCULATION

Seed should be treated with Emisan @ 2.5 g/kg before sowing. It helps in controlling the seed borne diseases (Bimbraw, 2013).

4.13 SEED RATE AND SOWING METHOD

For fodder production, required seed rate is 55–60 kg per hectare. Sow the seeds by pora or kera method in rows 25-cm apart at 3–5-cm depth. The crop can be sown with broadcast during the rainy season, if time is scarce. During rainy season when the moisture and humidity is high, the seed should be covered with thin layer of soil to ensure good germination (Bimbraw, 2013).

4.14 CROP MIXTURE

Sorghum can be sown in mixtures with cowpea, guar, soybean, pigeon pea, moth, black gram, valet bean, pillipesara, sunn hemp, etc.

4.15 CROP SEQUENCE

Some of the practicable crop sequences involving sorghum are as follows:

Sorghum + cowpea–oats + mustard–maize + cowpea

Sorghum + guar–berseem + mustard–maize + cowpea

Sorghum + guar–berseem + mustard–bajra + cowpea

Sorghum + guar–turnip–oats + mustard–maize + cowpea

Sorghum (M.P. Chari) 3 cuttings–berseem + mustard

Maize + cowpea–oats + mustard–sorghum (M. P. Chari) 2 cuttings

Sorghum + guar–lucerne

Sorghum + guar–senji/mustard (rainfed farming)

4.16 NUTRIENT MANAGEMENT (MANURES AND FERTILIZERS)

The use of farmyard manure proves to be useful in supplying phosphorus, potassium, and micronutrients to the crop, particularly under moisture stress conditions. Apply 8–10 t per hectare of farmyard manure in the field 20–30 days before sowing of the crop (Anonymous, 2009).

The addition of nitrogen on one hand increases the production of fodder and on other improves the quality by enhancing the protein content. The application of 80–100 kg N in two spilt in high rainfall or irrigated areas. Apply 40–50 kg of nitrogen (110 kg urea) and 20 kg P_2O_5 (125 kg single superphosphate) per hectare at sowing. Subsequently, second half of 40–50 kg of nitrogen (110 kg urea) per hectare is top dressed about 30 days after sowing (Singh et al., 2011).

In rainfed or low rainfall areas, drill 50 kg of nitrogen (110 kg urea) per hectare in between rows at sowing. Apply potassium to the crop on the soil test basis because some soils are deficient in phosphorous or potassium. In tropical areas, apply nitrogen, P_2O_5, and K_2O fertilizer at 60, 40, and 20 kg/ha, respectively (Bimbraw, 2013).

4.17 WATER MANAGEMENT

The water management varies with the season and soil type. The crop sown in March–April has the heavy requirement owing to high transpiration demand. Generally, five irrigations are sufficient to the March–June crop on medium soils. The number of irrigations has to be increased in light soils. For monsoon crop, one or two irrigations are sufficient depending upon rainfall. Drainage of excessive water during monsoon from the field is important for good crop.

4.18 WEED MANAGEMENT

At early stage the crop is very sensitive to weeds. For weed control, spray atrazine, at the rate of 1000 g/ha in 500 L of water, this should be done within 2 days of sowing. It provides control of annual weeds particularly chapatti (*Trianthema portulacastrum*). If guar is grown in mixture with sorghum, spray pendimethalin at the rate of 2.5 L/ha in 500 L of water within 2 days of sowing (Bimbraw, 2013; Mukherjee and Maiti, 2008).

In small areas, when sorghum is mixed with legumes, one or two intercultures with cultivators or hoes may be given. Some of the weed parasites such as *Striga*, cause severe damage to grain and fodder sorghums. These may be controlled by rotation of crops (Mukherjee and Maiti, 2008).

4.19 DISEASE AND INSECT-PEST MANAGEMENT

Some of the major diseases and insect pest, their causal organisms, symptoms, and control measures are given as follows:

4.19.1 DISEASES

4.19.1.1 SEED ROT AND SEEDLING MORTALITY

It is caused by seedborne fungi. It infects the seeds and seeds become shriveled and dark in color. Infected seedlings are damaged before they emerge out of soil and crops show poor stand. Control measure is to treat the seeds with Emisan @ 2.5 g/kg seed before sowing.

4.19.1.2 RED LEAF SPOT

It is caused by anthracnose and leaf blight, it affects the leaves. It spreads under high humidity. No chemical control is available for this disease, so it is recommended to grow resistant varieties. In summer, the occurrence of red leaf spots is not observed (Bimbraw, 2013; Mukherjee and Maiti, 2008).

4.19.1.3 GRAIN SMUT

It is caused by *Sphacelotheca sorghi*, it is a seed-borne disease. The fungus germinates with seed germination, grows inside the plant, and destroys all spikelets in the panicles. Seed treatment with copper sulfate or Thiram is effective against the fungus (Bimbraw, 2013; Mukherjee and Maiti, 2008).

4.19.1.4 DOWNY MILDEW

It is caused by *Sclerospora sorghi*, it attacks the whole plant. The plant turns grey and downy and eventually dies. Remedies are removal of diseased plant, crop rotation, and the use of resistant varieties (Mukherjee and Maiti, 2008).

4.19.2 INSECT-PESTS

4.19.2.1 SORGHUM SHOOT FLY

It is caused by *Atherigona varia soccata*. It damages the growing points and the bases of top leaves. The top leaf dies out and the growth of the shoot stops. To reduce the damage, timely sowing and insecticide application is required. Spraying of endosulfan and thimet at sowing are effective (Bimbraw, 2013; Mukherjee and Maiti, 2008).

4.19.2.2 STEM BORER

It is caused by *Chilo simplex*. The young caterpillars bore into clumps of the seedlings, kill the upper leaves, and growth is stopped. Systemic may be sprayed to control the pests. Application of carbofuran 3 g at 7.5 kg/ha in the whorl of plant or spraying endosulfan at 2 mL/L of water is sufficient for the control of stem borer (Bimbraw, 2013; Mukherjee and Maiti, 2008).

4.19.2.3 MITE

The leaves become red due to attack of mite. It may be controlled by spraying the crop with 625–1125 mL of malathion 50 EC in 250 L of water per hectare.

4.19.2.4 PYRILLA AND GREY WEEVIL

Spray 625–1250 mL of malathion 50 EC in 250 L of water per hectare on sorghum, depending upon the crop height if there is an attack of grasshopper, gray weevils, leaf hoppers, and pyrilla. For controlling pyrilla use the motorized knapsack sprayer (Bimbraw, 2013).

4.20 HARVESTING

Harvesting of fodder crop is done based on its use, at 50% flowering stage for making hay and at 50% flowering stage to milk stage for making silage and green feeding. Harvesting can be started from about 60 days, and continued up to 80 days, the crop remains safe at this stage due to absence of hydrogen cyanide (HCN) which is poisonous to animals. HCN toxicity can be reduced by applying irrigation 1 week before harvesting of crop, in drought conditions. Subsequent cuts are made after every 45–50 days (Bimbraw, 2013).

For multicut varieties such as Punjab Sudex chari 1, the first cutting is done on 55–65 days after sowing, and subsequent cuttings are to be taken after an interval of about 35–40 days (Bimbraw, 2013).

4.21 YIELD

Sorghum fodder crop yields about 500–600 quintal of green fodder per hectare depending upon the season. The crop yields 200–300 quintal per hectare green fodder under dry land areas (Bimbraw, 2013).

4.22 SEED PRODUCTION

On an average the seed yield of 10–12 quintal/ha can be obtained from the good crop. Seed production varies depending upon the quantity and distribution of rainfall. Good seed production is expected from the regions where the rainfall season is over early due to proper pollination and seed development (Bimbraw, 2013).

4.23 NUTRITIVE VALUE

The maximum nutrients in the fodder are available when the crop is cut at 50% flowering to milk stage. The quality of fodder crop also partly depends on the amount of fertilizer applied to the crop. Sorghum is suitable for hay and silage making. The HCN contents in the dried or ensiled sorghum reduce sharply and silage or hay presents no danger to animals. The crude protein content of sorghum is 7.75%; it may vary on the basis of stage of cutting and fertilizers applied. High quality forage is obtained from sorghum and legume mixtures under good fertilizer management and package of practices (Mukherjee and Maiti, 2008)

4.24 UTILIZATION

It is an ideal crop for silage and hay making, and can be utilized as green fodder under proper irrigation facilities (Bimbraw, 2013).

4.25 SPECIAL FEATURES TOXICITIES

HCN is present in early stages up to 40–50 days, so harvesting should be done at after 55–60 days of sowing and irrigation facilities should be taken care of, as in drought conditions the fodder crop contains more of HCN.

4.26 COMPATIBILITY

Sorghum can produce high quality forage when sorghum and legume mixtures are cultivated together under good fertilizer management and package of practices (Mukherjee and Maiti, 2008). It is also an ideal crop for silage and hay making, and can be utilized as green fodder (Bimbraw, 2013).

FIGURE 4.1 (See color insert.) Sorghum.

KEYWORDS

- **sorghum**
- **agronomic management**
- **nutritive value**
- **toxicities**
- **utilization**

REFERENCES

Anonymous. *Handbook of Agriculture*, 6th ed.; Directorate of Information and Publications of Agriculture, ICAR: New Delhi, India, 2009; pp 1054–1372.

Anonymous. *Nutritive Value of Commonly Available Feeds and Fodders in India*; National Dairy Development Board: Anand, India, 2012.

Bimbraw, A. S. *Production, Utilization and Conservation of Forage Crops in India*; Jaya Publishing House: Delhi, India, 2013.
Mukherjee, A. K.; Maiti, S. *Forage Crops Production and Conservation*; Kalyani Publishers: New Delhi, India, 2008; pp 44–86.
Panda, S. C. *Agronomy of Fodder and Forage Crops*; Kalyani Publishers: New Delhi, India, 2014; pp 13–16.
Singh, A. K.; Khan, M. A.; Subash, N.; Singh, K. M. *Forages and Fodder: Indian Perspective*; Daya Publishing House: New Delhi, India, 2011.

CHAPTER 5

JOHNSON GRASS (ALEPPO GRASS)

MD. HEDAYETULLAH[1*] and PARVEEN ZAMAN[2]

[1]*AICRP on Chickpea, Directorate of Research, Bidhan Chandra Krishi Viswavidyalaya, Kalyani, Nadia 741235, West Bengal, India*

[2]*Pulse & Oilseed Research Sub-station, Department of Agriculture, Government of West Bengal, Beldanga, Murshidabad 742133, West Bengal, India*

[*]*Corresponding author. E-mail: heaye.bckv@gmail.com*

ABSTRACT

Johnson grass is native to the Mediterranean region but is now found in essentially all temperate regions of the world. It was introduced to the United States in the early 1800s as a forage grass, and by the end of the 19th century Johnson grass was growing throughout the United States. In North Carolina, it occurs statewide. Each plant can produce hundreds of seeds with potential for far ranging dispersal by water, wind, livestock, commercial seed contamination, and contaminated machinery, grain, or hay. The immense, rapidly growing rhizome system of Johnson grass gives this plant a competitive edge allowing it to form dense colonies, displacing desirable vegetation, and restricting tree seedling establishment. In addition, the rhizomes regenerate easily from small pieces and are capable of growing or remaining dormant in a wide range of environmental conditions. Johnson grass occurs in temperate, subtropical, and tropical regions where it frequently occurs in ditches, field borders, cultivated lands, waste places, roadsides, other rights-of-ways, creeks, canal banks, and prairies.

5.1 BOTANICAL CLASSIFICATION

Kingdom: Plantae
Order: Poales
Family: Poaceae
Genus: *Sorghum*
Species: *halepense*
Binomial name: *Sorghum halepense* (L.) Pers.

5.2 COMMON NAME

It is also known as Aleppo grass, Aleppo millet grass, Barool (India), Baru grass (Pakistan), Arabian millet, Egyptian millet, Evergreen millet, Morocco millet, or Syrian grass

5.3 BOTANICAL NAME

Sorghum halepense (L) Pers.

5.4 INTRODUCTION

One of the most cosmopolitan of weeds, Johnson grass is thought to be native to the Mediterranean region but is now found in essentially all temperate regions of the world. It was introduced to the United States in the early 1800s as a forage grass, and by the end of the 19th century Johnson grass was growing throughout the United States. In North Carolina, it occurs statewide. Each plant can produce hundreds of seeds with potential for far ranging dispersal by water, wind, livestock, commercial seed contamination, and contaminated machinery, grain, or hay. The immense, rapidly growing rhizome system of Johnson grass gives this plant a competitive edge allowing it to form dense colonies, displacing desirable vegetation, and restricting tree seedling establishment. In addition, the rhizomes regenerate easily from small pieces and are capable of growing or remaining dormant in a wide range of environmental conditions. Johnson grass occurs in temperate, subtropical, and tropical regions where it frequently occurs in ditches, field borders, cultivated lands, waste places, roadsides, other rights-of-ways, creeks, canal banks, and prairies (Rout et al., 2013).

5.5 GENETICS OF THE PLANT

Johnson grass is about 8 ft tall, perennial, rhizomatous grass that grows in clumps or nearly solid stands. Stout stem that is pink to rusty-red near base. Smooth leaves with a characteristic white midvein. Appearing during the summer, flowers are large, loosely branched, and purplish. Seeds are reddish-brown and nearly 0.08 of an inch long Johnson grass has a chromosome number of either 2n = 20 or 2n = 40. The diploid races originated in South Eurasia. Populations from the United States and Canada are of Mediterranean origin and are a tetraploid ecotype. The tetraploid Johnson grass (2n = 40) will hybridize with diploid cultivated sorghum, *Sorghum bicolor* (L.) Moench (2n = 20), and the progeny are usually sterile triploids which may be weedy in subsequent crops. Occasionally fertile hybrids occur, one of which is believed to be the origin of the fodder crop Columbus grass or black Sudan grass, *S. almum* Parodi (2n = 40) in Argentina. The hybridization with sorghum is relatively rare and difficult to achieve (Dweikat, 2005).

5.6 PLANT CHARACTERISTICS

Johnson grass leaves are alternate. Blade—flat, smooth, 150–700-mm long × 5–30-mm wide, parallel sided, pointed tip, and obvious midrib on the lower surface. Edges of the leaf are white, rough to touch, and often wavy. Hairless except near the junction with the sheath which may be hairy rolled in the bud. Ligules are membranous and papery, 2–5-mm long, usually flat on top, jagged and fringed. Auricles are absent. Sheaths are loose, broad, striped, ribbed, and hairless.

5.6.1 STEMS

Stem are slender to stout, erect, tufted, spreading, 500–1500-mm tall, occasionally to 3000-mm tall. Stem is hairless apart from nodes which may be finely hairy. Lower nodes that are rooted are always hairy and sometimes bent at the nodes arising from an extensive, creeping, rooting, white, reddish or purple spotted, scaly rhizome.

5.6.2 FLOWER HEAD

Flower head are pale green to purplish, pyramidal panicle, 100–400-mm long × 50–250-mm wide and compact to open and spreading. The flower head is often hairy at the nodes. Many, slender branches are in rings or single, dividing into finer branches, bare at the base, jointed, angular, rough on the angles and hairy at the joints. Tips of the stalks are cup shaped. Spikelet clusters are held close to panicle branches, 10–25-mm long, fragile, with pale hairs on the stalks and joints.

5.6.3 FLOWERS

Spikelets are usually in pairs on branches and groups of three at the end. Flowers are of two types: stalked and stalkless. Stalkless ones are bisexual, jointed at the base, fall off easily, pale green or sometimes reddish, 4.5–5.5-mm long, almost pointed tip, awned or awnless, and glossy. Stalked spikelets are male or sterile, pale green to reddish or purplish, narrowly egg shaped, 5–7-mm long, and not shiny. Lower ones are reduced to an empty lemma and upper ones are bisexual in the stalkless spikelet. Glumes are on stalkless spikelet, leathery, stiff, narrowly egg shaped, all about the same size, 4.5–5.5-mm long × 2–3-mm wide. Lower glumes have 5–9 ribs near tip. Upper glume is three to seven nerved, hairy. On stalked spikelet, outer glumes, four times as long as wide, seven ribbed with a pointed tip.

5.6.4 SEEDS

Red brown to black, 3–4-mm long, egg shaped. Surface is finely lined.

5.6.5 ROOTS

The roots are creeping, rooting, scaly, branched, underground, purple spotted, rhizome to 300-mm deep. Fibrous roots extend up to 1200-mm deep and 1000 mm in diameter. A root is extended up to 3-m high with white and reddish scaly rhizomes. Leaf blades have a conspicuous midrib and sheaths are usually hairless and ribbed. Inflorescence is a purplish-brown, open pyramid shape with many slender main branches, dividing into finer branches, up to 20-cm by 12-cm wide, produced in summer.

5.7 ORIGIN AND DISTRIBUTION

Johnson grass is a native of the Mediterranean region but grows throughout Europe and the Middle East. The plant has been introduced to all continents except Antarctica, and larger islands and archipelagos. It reproduces by rhizome and seeds.

5.8 CLIMATIC REQUIREMENT

Johnson grass is grown in tropical, subtropical, and temperate areas. Monaghan (1980) has described a low-temperature suppression of rhizome bud sprouting, although some buds remain inactive even with a return to favorable growing conditions, ensuring that the plant can reestablish after temporary adverse conditions. Relatively high temperatures of 23–30°C promote bud sprouting. Fresh rhizomes are intolerant of high temperatures, being killed by 1–3 days on the soil surface at 50–60°C. Rhizomes will not tolerate very low soil temperatures, being killed at −17°C, but surviving −9°C (Stoller, 1977). Rhizome production and flowering are inhibited at temperatures below 13–15°C. *Johnson grass* will tolerate a degree of flooding, rhizome fragments germinating after being submerged for up to 4 weeks.

5.9 SOIL AND ITS PREPARATION

Johnson grass prefers moist soils. Johnson grass is mainly grown on low-potential, shallow soils with a high clay content, which usually are not suitable for the production of maize. Sorghum usually grows poorly on sandy soils, except where heavy textured subsoil is present. Sorghum is more tolerant of alkaline salts than other grain crops and can therefore be cultivated successfully on soils with a pH between 5.5 and 8.5. Sorghum can better tolerate short periods of water logging compared to maize. Soils with a clay percentage of between 10% and 30% are optimal for sorghum production.

5.10 VARIETIES

Local types are popularly grown for green fodder production.

5.11 SOWING TIME

Johnson grass can be planted between first week of May and first week of June. It is important to wait until soils are warm enough before seeding. The cool soil results the slow growth and gives weed seeds advantage over the crop. Late planting is not recommended, especially if it is grown for seed. Very hot summers or early fall rains reduce yields.

5.12 SEEDING METHOD

Johnson grass may be seeded either in cultivated rows or in close drills for hay purposes. In the drier sections better results will be obtained by seeding in cultivated rows having space of 18 in. or more apart. This method provides the individual plants with a greater area from which to receive sufficient moisture. Under high rainfall or irrigated conditions, use a row spacing of 6 in. or more. If the crop is being grazed by animals, use at least 20 in. spacing between rows to allow animals to move in the field without trampling the crop. Depending on available moisture and soil type, planting depths can range from 1 to 1.5 in. Cover seed with soil to enhance germination. Growers usually drill or broadcast Sudan grass at a rate of 11–15 kg of pure and live seeds per acre. Generally, lower seeding rates are better for dry areas. Higher rates are better for irrigated plantings or where rainfall is adequate. Research has shown that total forage yield is about the same regardless of the row spacing, because both Sudan grass and its hybrids tiller extensively.

5.13 SPACING

The row to row distance varies between 25 and 40 cm in moderate rainfall areas and 50 cm in dry areas.

5.14 SEED RATE

Seed rate varies from 15 to 75 kg/ha. The lower seed rate is maintained in dry condition and higher seed rate are maintained in humid conditions. The normal seed rate is 25–35 kg/ha.

5.15 SEED TREATMENTS AND INOCULATION

The seed should always be purchased from a reliable source. If seed is not already treated, seed may be treated with fungicides Thiram or agrosan GN @ 3 g/kg of seed. For nitrogen fixation, seed may inoculate with azotobacter @ 20 g/kg seed.

5.16 CROP SEQUENCE AND CROP MIXTURE

Johnson grass belongs to the family Poaceae (Graminae), therefore it is sown as a sole crop followed by leguminous fodder crops. In crop mixture, Sudan grass is grown together as a mixed cropping or intercropping system with leguminous fodder crops are given below:

i. Johnson grass–berseem + mustard
ii. Johnson grass–berseem + oats
iii. Johnson grass + cowpea
iv. Johnson grass–jower + cowpea
v. Johnson grass + cowpea
vi. Johnson grass + guar

5.17 MANURE AND FERTILIZER APPLICATION

During land preparation 10–20 t farm yard manure is recommended for 1 ha of land for higher green fodder yield. Use soil tests to determine the crop's nutrient needs. Fertilizer requirements are usually similar to those of other annual grass crops. Nitrogen (N) is the most limiting nutrient in Sudan grass production. Phosphorous moves slowly in the soil so apply it before planting or band it at seeding. Adjust the amount of fertilizer you need based on the previous crop. Adjust the rate and time of application of fertilizer to moisture supply and forage needs for optimum production. Split applications are important in order to reduce the possibility of nitrate poisoning in animals. Divide applications equally and apply after each grazing or harvesting operation. The recommended doses of nitrogen, phosphorus, and potash in kilogram per acre are 100–120: 25–30: 30–40, respectively, out of which 45–50 kg of nitrogen is applied at the time of land preparation. After each cut, 20 kg per acre more nitrogen should be applied to encourage growth.

5.18 IRRIGATION

When Johnson grass is used mostly as a forage crop it will be benefited under irrigation. Again, in sections where the precipitation is low and the supply of irrigation water is limited, much larger yields of forage will often be obtained by applying the small amount of water. When Johnson grass is planted early in the spring, it will be necessary to irrigate the ground before seeding, as the natural moisture will usually be sufficient to germinate the seed. Normally three or four irrigations throughout the growing season will be sufficient under most conditions. The first irrigation should be given when the plant reaches a height of 18–25 in. or just before heading. The later irrigations may be applied after each cutting or as soon as the new growth has started.

5.19 WEED MANAGEMENT

Johnson grass and weed competition is more, up to 60 days after sowing. Weed can suppress fodder crops if there is heavy infestation. Use cultivation methods if row spacing is adequate. Plant weed-free seed because weed problems in annual grasses most often come from the seed source. Buy seed from reputable dealers. Atrazine 1 kg a.i./ha preemergence helps also in controlling weeds.

5.20 INSECT-PEST AND DISEASE MANAGEMENT

Johnson grass affected by the leaf miner (*Hispellinus moestus*). Bacterial leaf spot is caused by *Pseudomonas syringae* and loose smut by *Sphacelotheca holci*, which is known to reduce plant growth and including rhizomes. Apply indoxacarb @ 1 mL per liter for controlling of leaf minor. Apply kasugamycin @ 400–500 mL per acre for effective control of bacterial leaf spot. Apply Mancozeb 75% WP @ 600–600 g per acre for controlling loose smut disease of Johnson grass.

5.21 HARVESTING

The first cut is usually taken about 90–100 days after planting or sowing. Clipping is done at 4–5 cm above the ground level. The new growth comes from the adventitious buds near the soil surface. In North India, four cuttings are taken while in the Southern and Eastern India five to six cuts are possible. Clipping may be done when the grass attains a height of 15–20 cm. After three to four times grazing, the grass should be clipped close to the ground to control weeds. The grass withstands trampling in grazing.

5.22 YIELD

The average green herbage yields from each cut are about 70–80 quintals/ha. The monsoon yields are higher than those of the hot and dry season. The average annual green herbage yields are 300 quintals/ha in North and 450 quintals/ha in South India.

5.23 SEED PRODUCTION

Fresh seed will not germinate for a few months after harvest. After this, 20–40% of seed is hard or dormant and will not germinate for some years. In the soil, it may remain viable for more than 10 years and has a half-life of just over 2 years. Dormancy is due to the persistent glumes and seed coat and if these are removed, nearly all seeds germinate including freshly harvested seed. Germination is generally higher on sands than clays. Seed will not germinate if the temperature is less than 15°C. Germination is greatest in light, alternating temperature, high nitrate, and moist conditions. There is considerable variation in germination characteristics between the varieties.

5.24 UTILIZATIONS

Johnson grass is an important fodder grass in many subtropical areas, having similar protein content to alfalfa and a comparable feed value to other important fodder grasses (Bennett, 1973; Looker, 1981). Due to its capacity to form extensive networks of rhizomes, Johnson grass can also be useful for control of soil erosion (Bennett, 1973). Johnson grass has been used for forage and to stop soil erosion, but it is often considered as a weed. The foliage can cause "bloat" in such herbivores from the accumulation of excessive nitrites; otherwise, it is edible. It grows and spreads so quickly, it can "choke out" other cash crops planted by farmers. These species occur in crop fields, pastures, abandoned fields, rights-of-way, forest edges, and along stream banks. It thrives in open, disturbed, rich, bottom ground, particularly in cultivated fields. Johnson grass resistant to the common herbicide glyphosate has been found in Argentina and the USA. It is considered to be one of the 10 worst weeds in the world. It is named after an Alabama plantation owner, Colonel William Johnson, who sowed its seeds on river bottom farmland circa 1840. The plant was already established in several US states a decade earlier, having been introduced as prospective forage or accidentally as a seed lot contaminant.

5.25 SPECIAL FEATURES TOXICITIES

It is toxic when stressed, frosted, wilted, or stunted. It produces hydrogen cyanide (HCN) and causes cyanide poisoning. Young shoots are probably the most toxic. Cattle, sheep, and pigs have been poisoned in Australia. Cattle appear to have suffered the worst. It may also contain toxic levels of nitrate and cause nitrate poisoning especially during periods of vigorous growth. New growth of Johnson grass produces cyanide, which can build up to levels dangerous to stock in summer. It is a rhizome perennial which grows rapidly in favorable conditions and forms dense persistent stands, but is restricted to habitats where water is abundant during the summer growing season; various sources indicate that, although Johnson grass has been widely used as a fodder plant, it can cause poisoning of cattle under some circumstances due to its cyanic content (Henderson, 2001) during periods of vigorous growth, drought, or following frost. Consequently, prussic acid poisoning of cattle is well known in Australia and USA.

KEYWORDS

- **Johnson grass**
- **cultivation practices**
- **yield**
- **nutritive value**
- **utilization**
- **toxicities**

REFERENCES

Bennett, H. W. Johnson Grass, Carpet Grass and Other Grasses for the Humid South. In *Forages*; Heath, M. E., Metcalf, D. S., Barnes, R. F., Eds.; Iowa State University Press: Ames, USA, 1973; pp 286–293.

Dweikat, I. A Diploid Interspecific, Fertile Hybrid from Cultivated Sorghum, Sorghum Bicolor, and the Common Johnson Grass Weed, *Sorghum Halepense. Mol. Breed.* **2005,** *16* (2), 93–101.

Henderson, L. Alien Weeds and Invasive Plants. In *Plant Protection Research Institute Handbook No. 12*; Paarl Printers: Cape Town, South Africa, 2001.

Looker, D. Johnson Grass has an Achilles Heel. *New Farm.* **1981,** *3*, 40–47.

Monaghan, N. The Biology of Johnson Grass (*Sorghum Halepense*). *Weed Res.* **1980,** *19* (4), 261–267.

Rout, M. E.; Chrzanowski T. H.; Smith, W. K.; Gough, L. Ecological Impacts of the Invasive Grass *Sorghum Halepense* on Native Tallgrass Prairie. *Biol. Invasions* **2013,** *15* (2), 327–339. (2013a.)

Stoller, E. W. Differential Cold Tolerance of Quack Grass and Johnson Grass Rhizomes. *Weed Sci.* **1977,** *25* (4), 348–351.

CHAPTER 6

JOB'S TEARS *(COIX)*

R. PODDAR[1], K. JANA[1], CHAMPAK KUMAR KUNDU[1], and H. DAS[2]

[1]*Department of Agronomy, Faculty of Agriculture, Bidhan Chandra Krishi Viswavidyalaya, Mohanpur 741252, Nadia, West Bengal, India*

[2]*GKMS, Odisha University of Agriculture and Technology, Kalimela, Malkangiri, Odisha, India*

[*]*Corresponding author. E-mail: rpoddar.bckv@rediffmail.com*

ABSTRACT

Job's tears is a group of grasses belonging to the family Poaceae and popularly known as "Job's tears." Under that designation is included not merely the species of *Coix* but of *Chionachne*, and probably also of *Polycota.* The plant is a native of Southeast Asia and is found widely distributed throughout the tropical and subtropical parts of the world. Indication is given of an ancient cultivation which may have taken its birth in China and spread through the Malayan regions into Burma and then to Assam and to other places in India. On an average the coix green fodder yielded about 600–700 quintal per hectare under good management practice. Coix seed yield varies from 1.8 to 2.1 q ha^{-1} depending upon crop management. The herbage contains 8.5% crude protein and 28.0% crude fiber. It is mixed with molasses and concentrates to make it more palatability.

6.1 BOTANICAL CLASSIFICATION

Kingdom: Plantae
Order: Poales
Family: Gramineae (Poaceae)

Subfamily: Panicoideae
Tribe: Andropogoneae
Genus: *Coix*
Species: *lacryma-jobi*

Genus of *Coix* Linn. of about eight to nine species under the family Gramineae (Poaceae) of which three to four species are more common as cultivated crop. These are:

- *Coix lacryma-jobi* var. *lacryma-jobi*. Widely distributed throughout the Asian subcontinent to Peninsular Malaysia and Taiwan; naturalized elsewhere.
- *Coix lacryma-jobi* var. *ma-yuen* (Rom. Caill.) Stapf. South China to Peninsular Malaysia and the Philippines.
- *Coix lacryma-jobi* var. *puellarum* (Balansa) A. Camus. Assam to Yunnan (China) and Indochina.
- *Coix lacryma-jobi* var. *stenocarpa* Oliv. Eastern Himalayas to Indochina.

6.2 COMMON NAME

Job's tears is also known as coix seed, tear grass, hato mugi, adlay, or adlai.

6.3 BOTANICAL NAME

Coix lacryma-jobi.

6.4 INTRODUCTION

Job's tears is a group of grasses belonging to the tribe Maydeae and popularly known as "Job's tears." Under that designation is included not merely the species of *Coix* but of *Chionachne*, and probably also of *Polycota.* The popular or rather practical view of these plants will be adopted in the following brief description of these species of "Job's Tears" (Watt, 1972). The plant is a native of Southeast Asia and is found widely distributed throughout the tropical and subtropical parts of the world. Indication is given of an ancient cultivation which may have taken its birth in China and spread through the Malayan regions into Burma and then to Assam and to other places in India.

6.5 BOTANICAL DESCRIPTION

It is robust, erect, tall, with brace roots arising from the lower nodes. The grass is monoecious with separate male and female flowers. It is annual or perennial grass. Generally plant height is 1–2 m to its maximum of 3 m., stems are branching, spongy within the internodal portion. Leaves are long, flat, and broad enough. It bears large, shining, pear-shaped fruits showing a fanciful resemblance to tears and containing a whitish or light brownish grain similar to rice. Inflorescences are formed in the axils of upper leaves on simple peduncles of 3–6-cm long. At the end of each peduncle, hollow, and globular beadlike structures are developed, enclosing pistillate spikelets. Seeds are yellow, purple, white, or brown. There are soft-shelled forms for eating and hard-shelled forms for making ornaments, especially necklaces. The wild forms yield fruits with smooth and brilliantly shining shells which may be grey, bluish grey, brown, or black in color. Test weight is (100 seed) 16–17 g. It is a medium stony seed. Flowering is nonsynchronous, seed highly shattering. Plant matures in 150–160 days.

6.6 ORIGIN AND DISTRIBUTION

The plant is a native of Southeast Asia and is found widely distributed throughout the tropical and subtropical parts of the world. Indication is given of an ancient cultivation which may have taken its birth in China and spread through the Malayan regions into Burma and then to Assam and to other places in India generally (Watt, 1972). It is cultivated in the tropics, particularly in the Philippines as an auxiliary food crop. In India, it is met within the plains. It is also cultivated in African countries and in America. Now, "adlay" is cultivated extensively in the Philippines, Indochina, Siam and Burma, and to some extent in Sri Lanka.

6.7 PLANT CHARACTERISTICS

It is a tall, aquatic grass with large broad leaves, found throughout the plains of India. It is wild in low-lying areas of West Bengal, Bihar, and Assam as a floating weed. It is very varying in habit and characteristics and several races show marked differences in the size, shape, color, and degree of the hardness of the outer coat of the fruits. Inflorescences are formed in the axils of upper leaves on simple peduncles of 3–6-cm long. At the end of the peduncle, hollow, and globular beadlike structures develop, enclosing pistillate spikelets. Seeds are yellow, purple, white, or brown in color. There are soft-shelled forms for eating and hard shelled ones for making ornaments specially necklaces (Thomas, 2003).

6.8 CLIMATIC REQUIREMENT

Although the crop may be grown as perennial but the main growing period is rainy season, in marshy low-lying areas. Hot, humid accomplished with sufficient moisture in the soil, is the ideal condition for growth and development of the crop (Puste, 2004).

6.9 SOIL AND ITS PREPARATION

It can be grown on all types of soil, preferably clay to clay loam soil. It can also be grown in marshy-swampy soils. It can be tolerant to salinity. Some species can grow in warm slopes of hills up to 1600–1700 m above sea level.

6.10 VARIETIES

6.10.1 BIDHAN COIX 1

This variety has been developed by Bidhan Chandra Krishi Viswavidyalaya, Kalyani, and released for cultivation in West Bengal, Odisha, Assam, and North Bihar. The average green fodder, dry matter, and crude protein yield is 34.6 t/ha, 6.9 t/ha and 0.5 t/ha, respectively. [CVRC Notification no. 449 (E) dated February 11, 2009].

6.10.2 KCA-3

This variety was developed by BCKV, Kalyani in 2004 and is recommended for cultivation in states of Assam, Odisha, Jharkhand, West Bengal, Meghalaya, and Bihar.

6.10.3 KCA-4

This variety was developed by BCKV, Kalyani in 2005 and is recommended for cultivation in Assam, Odisha, Jharkhand, West Bengal, Meghalaya, and Bihar.

6.11 LAND PREPARATION

Lands should be sufficiently plowed and leveled enough by repeated laddering operations during kharif season. Presence of moisture in the soil is most important and should be sufficient for good germination. From germination to initial growth stages there is no need of standing water. But moist soil is required for growth and development at later stages. Seeds are used as propagating materials for the cultivation of *Coix* spp. Although it can grow on well-drained highlands as well as its cultivation is confined to low-lying marshy wetlands.

6.12 SOWING TIME

Sowing can be done anytime of the year except winter month.

6.13 SEED RATE AND SOWING METHOD

There are two types of sowing methods generally practiced, namely, broadcasting and line sowing. Common practice is the broadcasting of seeds in the field. It can be sown in line also. Seed rate is 30–40 kg ha^{-1}. Deep plowing followed by laddering is required for good establishment of the crop. The space between rows and between plants should be kept as 40 cm × 15 cm. Seeds are also placed at a depth of 5–7 cm. Seedlings of 1month of age are transplanted in the water stagnant areas for better establishment of the crop. The seed coat of coix is very hard, for that reason seeds should be soaked for 24 h before sowing.

Sufficient rains in the early stage of the growth and comparatively a dry period, when the grain is setting—if it is considered as a grain crop—are necessary for good yield.

However, in most cases it is harvested as a green succulent fodder. It may be a good fodder crop, particularly during rainy months when all other common fodder crops are not really available to a farmer.

6.14 NUTRIENT MANAGEMENT

The plants respond well to the liberal application of organic manures. Generally farmers do not apply any manures and fertilizers to the crop. However, organic manure either farm yard manure or well-rotten cow dung @ 10–12 t ha^{-1} should be applied. The crop requires NPK @ 80:40:40 kg ha^{-1} for better growth. 50% of N and the entire P and K are applied at the time of sowing before final land preparation as basal and the remaining 50% of nitrogen is applied in equal splits after each cutting. It will be sufficient for a good crop stand which favors more tonnage of production of green fodder. At every application, fertilizers should be mixed well with the soil so that it will reduce the leaching loss of nitrogen. This balance application of nutrients will accelerate the growth of the crop as well as will produce more tonnage of green forage.

6.15 WEED MANAGEMENT

One hand hoeing or hand weeding at 25 days after sowing (DAS) is enough for weed management. Some of the common marshy aquatic weeds may compete with *Coix*, they should be eradicated from the field, for this two to three times weeding at 20–25 and 40–45DAS is necessary. Soil stirring at the time of weeding is helpful for better growth, development of roots, initiation of tillering, etc.

6.16 IRRIGATION MANAGEMENT

Irrigation may be given during summer and postmonsoon period at an interval of 15–20 days.

6.17 PESTS AND DISEASES

No important pest is recorded. Leaf spot, tar spot, leaf rust, etc. are important diseases of *Coix*. But losses may occur due to depredations of rats and parrots (Anonymous, 1939). It is said to be liable to smut attack.

6.18 CROPPING SYSTEM

Job's tear may be grown in a mixed cropping system. It can be mixed with rice bean and rice crop for better yield.

6.19 HARVESTING

First cutting was given at an age of 80 days after planting and subsequent cutting is given at an interval of 4–6 weeks depending upon the growth of crop. During cutting 30 cm stubble above ground level is left. Two to three cuttings may be taken per year.

6.20 YIELD

On an average the coix green fodder yielded about 600–700 quintal per hectare under good management practice.

6.21 SEED PRODUCTION

If coix is cultivated for seed purpose then seed yield varies from 1.8 to 2.1 q ha^{-1} depending upon crop management.

6.22 NUTRITIVE VALUE

Nutritive value of coix grain is albuminoids 18.2%, starch 58.3 %, oil 5.2%, and fiber 1.5%. As per reports, "*adlay*" compares favorably with other grains as regards protein and fat contents, but the mineral content is low. The phosphorus content is high and the calcium content is low. A prolamins, coixin is rich in leucine and glutamic acid, has been isolated from the grains (Anonymous, 1988).

6.23 UTILIZATION

It may be grown as a fodder plant in marshy areas which cannot be easily drained and which retain but insufficient water for transplanted paddy crops. The foliage of young plants is used as fodder. In latter state, cattle generally do not prefer owing to its course in nature. However, it is said to be used as a fodder crop in absence of other green forage during this time. The herbage contains 8.5% crude protein and 28.0% crude fiber. It is mixed with molasses and concentrates to make it more palatable. On breaking the outer shell, a cowry-shaped grain is obtained. The grains can be used after grinding as poultry feed. It can also be used as silage. It can be used in the preparation of any article of food that is usually made of rice and with the same degree of palatability. Perched grains are made into beverage in Japan. A light beer, "*Dzu*" is made from it by the "Nagas." The fruits after hulling can be fed to poultry (Anonymous, 1988).

6.23.1 MEDICINAL VALUE

The grains are considered a good blood purifier and excellent diuretic. The fruits are used in medicine either as tincture or as decoction, for catarrhal affection of the air passage and inflammation of the urinary passage. The kernels deprived of their shells are used as food and medicines throughout China, Malaya, Indochina, Philippines, and in La Reunion. They make an excellent diet drink.

6.23.2 DOMESTIC USES

The foliage may be used as fodder for cattle, horses, and elephants. The native straw and leaves may be used for thatch. In many localities the wild, hard, dry, spherical grain is extensively used by the aboriginal races for ornamental purposes. Necklaces of these seeds are frequently worn, and baskets and other ornamental articles are occasionally decorated with them.

FIGURE 6.1 (See color insert.) Coix.

KEYWORDS

- *coix*
- **cultivation practices**
- **yield**
- **utilization**
- **nutritive value**

REFERENCES

Anonymous. *Trop. Agriculturist* **1939,** *93,* 352.

Watt, G. *A Dictionary of the Economic Products of India*; Cosmo Publications: New Delhi, 1972; Vol. II, pp 491–500.

Anonymous. *The Wealth of India, Vol. II*; CSIR: New Delhi, 1988; pp 305–306.
Thomas, C. G. *Forage Crop Production in the Tropics*; Kalyani Publishers: New Delhi, 2003; pp 96–97.
Puste, A. M. Aquatic Non-food Commercial Crops. In *Agronomic Management of Wetland Crops*; Kalyani Publishers: New Delhi, 2004; pp 248–250.

CHAPTER 7

DEENANATH GRASS *(PENNISETUM)*

MD. HEDAYETULLAH[1*] and PINTOO BANDOPADHYAY[2]

[1]*AICRP on Chickpea, Directorate of Research, Bidhan Chandra Krishi Viswavidyalaya, Kalyani 741235, Nadia, West Bengal, India*

[2]*Department of Agronomy, Bidhan Chandra Krishi Viswavidyalaya, Mohanpur, West Bengal, India*

[*]*Corresponding author. E-mail: heaye.bckv@gmail.com*

ABSTRACT

The meaning of the Deenanath grass is friend of the poor. *Pennisetum* is a native of Africa and India. Deenanath grass also known as desho or as desho grass, is an indigenous grass of Ethiopia belonging to the Poaceae family. The genus *Pennisetum* is derived from the Latin word "penna," feather, and "seta," bristle and the species pedicellatum means having small stalk of spikelets. Desho is becoming increasingly utilized, along with various soil and water conservation techniques, as a local method of improving grazing land management. The average yield from single cut is about 400–500 quintals/ha of green fodder and 40–60 quintals/ha dry fodder. Green fodder yield under irrigated fodder crop is about 400–600 quintals/ha first crop and ratoon crop gives about 150–200 quintals/ha. Desho is used as a year round fodder and it has no toxic effect in animals and livestock.

7.1 BOTANICAL CLASSIFICATION

Kingdom: Plantae
Order: Poales
Family: Poaceae

Genus: *Pennisetum*
Species: *pedicellatum*
Binomial name: *Pennisetum pedicellatum* Trin.

7.2 COMMON NAME

Deenanath grass, desho grass, Deenabandhu grass, Deena, Dina in India, kyasuwa in Nigeria, bare in Mauritania.

7.3 BOTANICAL NAME

Pennisetum pedicellatum Trin.

7.4 INTRODUCTION

Pennisetum pedicellatum, known simply as desho or desho grass, is an indigenous grass of Ethiopia belonging to the Poaceae family of monocot angiosperm plants. The genus *Pennisetum* is derived from the Latin word "penna," feather, and "seta," bristle and the species *pedicellatum* means having small stalk of spikelets. It is also known as annual kyasuwa grass in Nigeria, bare in Mauritania, and Deenanath grass in India. The meaning of the Deenanath grass is friend of the poor. It grows in its native geographic location, naturally spreading across the escarpment of the Ethiopian highlands. Widely available in this location, it is ideal for livestock feed and can be sustainably cultivated on small plots of land. Thus, desho is becoming increasingly utilized, along with various soil and water conservation techniques, as a local method of improving grazing land management and combating a growing productivity problem of the local region.

7.5 PLANT CHARACTERISTICS

Deenanath is an herbaceous perennial grass which has a massive root system that anchors to the soil. It has a high biomass producing capacity and grows upright with the potential of reaching 90–120 cm and often produces 30–60 tillers per plant. The culms are tinged reddish-purple at their base. The leaves are 40–50-cm long and 2–2.5-cm broad. The inflorescence is a spikelike panicle, 15–20-cm long. The seeds are small and shiny. The grass is very leafy and provides forage for longer periods than sorghum and bajra. The leaf:stem ratio of the herbage is 1:1.22, whereas in sorghum it is 1:2. Initial growth of this grass is slow and after 10–12 weeks, the growth is fast reaching a maximum growth at 16 weeks. The plant height depends on soil fertility (Shiferaw et al., 2011). Desho is planted by cuts which have good survival rates and establishes better compared to grasses planted by seed. Moreover, desho grows rapidly and is drought resistant once established. Desho is said to have high nutritive values and is naturally palatable for livestock.

7.6 ORIGIN AND DISTRIBUTION

Pennisetum is a native of Africa and India. The plant has been introduced to all continents except Antarctica, and larger islands and archipelagos. It reproduces by rhizome and seeds. *Pennisetum* naturally occurs in Bihar, West Bengal, Odisha, Madhya Pradesh, Andhra Pradesh, and Tamil Nadu.

7.7 CLIMATIC REQUIREMENT

Pennisetum can be grown throughout the year under irrigated condition. Optimum sowing time is monsoon season. The grass prefers a warm climate. It thrives well in dry and moderately humid areas with rainfall ranging from low (50 cm) to high (150 cm). It grows well in 30 cm rainfall supplemented by two to three irrigations. It particularly prefers a good moist soil during its active growth period. It can adapt well in even on eroded marginal wastelands, bank of marshy land, and poor soil but gives much higher yields on fertile well drained soil. The grass is not suitable in flood-prone areas and in the cold in high Himalayan regions. The grass can withstand in acidic as well as alkaline soils. It was found to be most tolerant to soil acidity as it could be grown without any application of lime. It could tolerate soil salinity to the tune of 12 mmhos/cm. It safely grows with pH ranges 5.5–8.0 (Chatterjee and Das, 1989).

7.8 SOIL AND ITS PREPARATION

All types of soil with good drainage. Deenanath grass does not come up well on heavy clay soil or flooded or waterlogged conditions. Plow two to three times by desi plow or blade harrow to a depth of 10–12 cm to obtain good tilth and form beds and channels. The field does not require deep plowing and repeated plowing. Two to three times crosswise discing or harrowing is also better for cutting of grassy weeds. The field should be free from weed for better establishment.

7.9 VARIETIES

Some promising varieties of *Pennisetum* are type-3, type-10, type 15, CO-1, Bundle Dinanath-2 (NC), and Bundle Dinanath-1 (NC), Jawahar Pennisetum 12, T 3, T4, T 5, T 10, T 38, T 42, IGFRI 3808, Pusa Deenanath, IGFRI 966-1, Bundel 1, Bundel 2, TNDN 1, IGFRI 4-2-1, and IGFRI 43-1.

7.10 SOWING TIME

The seed is generally sown in June and July months. It may also be sown in August to get fodder by the month of November. Transplanted Pennisetum may also sow in May–June for raising seedlings and finally transplanted in the main field after receiving the monsoon rains. In irrigated crop, seed is sown in early March for getting 2–3 cuttings.

7.11 SEEDING METHOD

The seeds are fine and fluffy. Seed is sown by broadcasting hampering due to the fluffiness of seed coat but it can also be sown in line. The seed is usually mixed with moist soil or powered soil or fine sand for uniform broadcasting of the seed. The seeding depth should be 1–2 cm below the soil surface. Shallow placement of seed gets better germination.

7.12 SPACING

The row to row distance varies between 25 and 30 cm in assured irrigated areas. The transplanted fodder crops may be spaced at 30 cm × 15 cm. The optimum plant population is maintained about 30–40 plant/m^2. Cuts of the grass are ideally planted in rows, spaced at 10 cm × 10 cm, using a hand hoe. This spacing gives each plant sufficient soil nutrients and access to sunlight to achieve optimal growth, while ensuring the soil will be completely covered by the grass once established.

7.13 SEED RATE

The seed rate is 5–7.5 kg/ha seed kernels in 20 cm apart in line sowing and 40 kg/ha with fluffs.

7.14 SEED TREATMENTS AND INOCULATION

The seed should always be purchased from a reliable source. If seed is not already treated, seed may be treated with fungicides Thiram or agrosan GN @ 3 g/kg of seed. For nitrogen fixation, seed may inoculate with azotobacter @ 20 g/kg of seed.

7.15 CROP SEQUENCE AND CROP MIXTURE

Deenanath grass belongs to the family Poaceae (Graminae), therefore it is sown as a sole crop followed by leguminous fodder crops or sown in mixture for better and quality fodder production. Initial growth of legumes is faster than the grass. Quality fodder productions are obtained in mix components. The technical specifications for cultivating pennisetum are essential to improve grazing land management practices.

It is recommended to plant other species alongside desho to promote biodiversity. Multipurpose shrubs trees, for example *Leucaena* sp. and *Sesbania* sp., can be planted approximately 5 m apart. Other legumes, such as alfalfa and clover, can be mixed in with desho by being broadcast throughout the plot. Once planted, maintenance activities such as applying fertilizer, weeding, and gap filling, ensure proper establishment and persistence of desho. Fertilizer should be applied throughout the plot 1 month after planting. Weeding and gap filling are continuous activities. In crop mixture, Deenanath grass is grown together as a crop sequence, mixed cropping, or intercropping system with leguminous fodder crops.

i. *Pennisetum*–berseem
ii. *Pennisetum*–lablab bean
iii. *Pennisetum* + cowpea
iv. *Pennisetum* + rice bean
v. *Pennisetum* + black gram
vi. *Pennisetum* + green gram or soybean

7.16 MANURE AND FERTILIZER APPLICATION

During land preparation or before 15 days of land preparation well decomposed FYM or compost is applied as basal @ 10–15 t/ha. Chemical fertilizers application during final land preparation as basal @ 80–100:60–120:40 kg/ha of NPK. Deenanath grass is highly responsive to nitrogenous fertilizers. Application of 50 kg N from FYM and 45 kg N from fertilizer gave 93 quintals/ha dry forage in a single cut. In sole fodder crop 80–100 kg N/ha may be applied (half basal and half top dressed after 6–8 weeks of growth). Application of 40 kg/ha N is conducive for better regrowth of the crop. Superphosphate applied @ 60 kg/ha for optimum fodder production as a basal.

The recommended dose of muriate of potash is 40 kg/ha for higher green fodder production. Top dressing at the rate of 20 kg N on 30th day after sowing, 50% of this has to be applied for rainfed crop. It is recommended to use organic compost in the form of animal manure, leaf litter, wood ash, food scraps, and/or any other rich biodegradable matters. After this initial treatment, fertilizer is only applied sporadically when desho plants are struggling to grow or where replanting has taken place.

7.17 IRRIGATION

Irrigation may be necessary every 2–3 weeks during postmonsoon depending upon the soil moisture and soil type. During rainy season, the field should be well drained to avoid lodging.

7.18 WEED MANAGEMENT

Initial crop growth not first, crop weed competition is more up to 60 days after sowing. Weed can suppress fodder crops if there is heavy infestation. Use cultivation methods if row spacing is adequate. Plant weed-free seeds, because weed problems in annual grasses most often come from the seed source. Buy seed from reputable dealers. Atrazine 1 kg a.i./ha preemergence helps also in controlling weeds.

7.19 INSECT-PEST AND DISEASE MANAGEMENT

The grass has not yet posed any serious problem for its cultivation in respect of pests and diseases.

7.20 HARVESTING

March sowing crop can be harvested (first cut) after 70–100 days and subsequent cut coincides with the flowering stage. Clipping should be done about 10 cm above the ground level for good regeneration. Sometimes last cut is allowed for seed setting. The usual interval between cuts is approximately 40–60 days. The severe summer and winter restrict the crop growth. The first cut is usually taken about 90–100 days after planting or sowing.

Clipping is done at 4–5 cm above the ground level. The new growth comes from the adventitious buds near the soil surface. In North India, four cuttings are taken while in the Southern and Eastern India five to six cuts are possible. Clipping may be done when the grass attains a height of 15–20 cm. After three to four times grazing, the grass should be clipped close to the ground to control weeds. The grass withstands trampling in grazing.

7.21 GREEN FODDER YIELD

The average yield from single cut is about 400–500 quintals/ha of green fodder and 40–60 quintals/ha of dry fodder. Green fodder yield under irrigated fodder crop is about 400–600 quintals/ha of first crop and ratoon crop gives about 150–200 quintals/ha. Application of high N dose at the rate of 120 kg/ha gives green fodder yield about 700–800 quintals/ha and 80–90 quintals/ha of dry matter.

The regrowth herbage from an early dry season crop with 80 kg N/ha can be as high as 1200 quintals/ha green forage or 260 quintals/ha dry matter. Yield potentiality of summer crop with irrigation is higher than that of rainy season crop. The average green herbage yields from each cut are about 70–80 quintals/ha. The monsoon yields are higher than those of the hot and dry season. The average annual green herbage yields are 300 quintals/ha in North and 450 quintals/ha in South India.

7.22 CONSERVATION

High quality hay is obtained from Deenanath grass. The grass is harvested at the boot stage, dried in thin layers in the sun for a day or two, and then in shade of trees to preserve the color and avoid decomposition. For silage, harvested fodder is wilted up to 65% moisture content and if possible mixed with cow pea, guar, berseem, lucerne, moong, moth, rice bean, or horse gram before ensiling.

7.23 NUTRITIVE VALUE

The forage crop is leafy, succulent, and palatable. The chemical composition of the grass in percentage for dry matter at flowering stage is crude protein 6.5%, ether extract 3.2%, crude fiber 35.8%, nitrogen free extract 40.1%, total acid 14.4%, organic matter 85.7%, calcium (Ca) 0.4%, and phosphorus (P) 0.3%.

7.24 SEED PRODUCTION

7.24.1 LAND REQUIREMENT

Land should be free of volunteer plants. The previous crop should not be the same variety or other varieties of the same crop. It can be the same variety if it is certified as per the procedures of the certification agency.

7.24.2 ISOLATION

For certified/quality seed production, leave a distance of 10 m all around the field from the same and other varieties of the crop.

7.24.3 PRESOWING SEED MANAGEMENT

Break the seed dormancy by mechanical scarification in a defluffer followed by soaking in a mixture of GA_3 (200 ppm) and KNO_3 (0.25%) (1:1) for 16 h. Pellet the seed with DAP @ 60 g/kg and arappu leaf (*Albizia amara*) powder @ 500 g/kg/ha of seed to enable easy handling of seed during sowing and also for better establishment.

7.24.4 SEED TREATMENT

Slurry treats the seeds with carbendazim @ 4 g kg^{-1} of seed.

7.24.5 HARVEST

Deenanath grass (CV Pusa 3) attained physiological maturity at fifth week after 50% flowering, while it took 6 weeks for cv. TNDN.1. Germination is higher for the seeds from the first and second formed tillers. Delayed harvesting resulted in shattering loss. The middle and proximal portions of the spike produce high-quality seeds.

7.24.6 STORAGE

Store the seeds in gunny or cloth bags for short-term storage (8–9 months) with seed moisture content of 8–10%. Store the seeds in polylined gunny bag for medium-term storage (12–15 months) with seed moisture content of 8–9%. Store the seeds in 700 gauge polythene bag for long-term storage (more than 15 months) with seed moisture content of less than 8% of harvest.

7.25 UTILIZATIONS

Desho is used as a year round fodder. To maintain the sustainability of the intervention, the plot is permanently made inaccessible to free-grazing livestock; instead a cut-and-carry system is encouraged. Cut-and-carry means that desho is harvested and brought to livestock for stall-feeding. Due to its rapid growth rate, desho provides regular harvests, even reaching monthly cuts during the rainy reason. Once a year, just before the dry season, sufficient grass is harvested and stored as hay to feed the livestock until the rains return (Danano, 2007). Deenanath grass can protect against runoff and soil loss on the slopes. This grass strips reduce soil loss by approximately 45% in the first few years of establishment compared to areas with no barriers. However, vetiver grass was found to be more effective than desho, and thus vetiver should be used as the preferred grass for hedgerow technology (Welle et al., 2006).

7.26 SPECIAL FEATURES (TOXICITIES)

No toxic effect has been found. The oxalic acid content is low (1.7%) as compared to pearl millet (2.5%).

KEYWORDS

- **Deenanath grass**
- **cultivation practices**
- **fodder yield**
- **nutritive value**
- **utilization**

REFERENCES

Chatterjee, B. N.; Das, P. K.; *Forage Crop Production Principles and Practices;* Oxford & IBH Publishing Co. Pvt. Ltd.: Kolkata, 1989; 254–262.

Danano, D. Improved Grazing Land Management: Ethiopia. In *Where the Land is Greener*; Liniger, H., Critchley, W., Eds.; WOCAT: Bern, Switzerland, 2007; pp 313–316.

Shiferaw, A.; Puskur, R.; Tegegne, A.; Hoekstra, D. Innovation in Forage Development: Empirical Evidence from Alaba Special District, Southern Ethiopia. *Develop. Pract.* **2011,** *21* (8), 1138–1152.

Welle, S.; Chantawarangul, K.; Nontananandh, S.; Jantawat, S. Effectiveness of Grass Strips as Barriers Against Runoff and Soil Loss in Jijiga Area, Northern Part of Somalia Region, Ethiopia. Kasetsart. *Nat. Sci.* **2006,** *40,* 549–558.

CHAPTER 8

TEOSINTE (MAKCHARI)

M. RANA[1], R. GAJGHATE[1], G. GULERIA[2], R. P. SAH[3], ANJANI KUMAR[3*], and O. N. SINGH[3]

[1]*Crop Production Division, Indian Council of Agriculture Research-Indian Grass and Fodder Research Institute, Jhansi 284003, Uttar Pradesh, India*

[2]*Department of Agronomy, Chaudhary Sarwan Kumar Himachal Pradesh Krishi Vishvavidyalaya, Palampur 176062, Himachal Pradesh, India*

[3]*Crop Production Division, Indian Council of Agriculture Research National Rice Research Institute, Cuttack 753006, Odisha, India*

[*]*Corresponding author. E-mail: anjaniias@gmail.com*

ABSTRACT

Teosinte name is commonly called for a group of four annual and perennial species of the genus *Zea*. Teosinte, the closest relative of maize is used extensively for green fodder but stover, silage, and hay is also used for livestock. Teosinte is used as feed and fodder for animals due to its vigorous growth, high yield, and tolerance to biotic and abiotic stresses. The quality of fodder is comparable to maize and possesses high percentage of carbohydrate.

8.1 COMMON NAME

Makchari, buffalo grass, or teosinte grass.

8.2 BOTANICAL NAME AND CLASSIFICATION

Kingdom: Plantae
Subkingdom: Tracheobionta
Super division: Spermatophyta
Division: Magnoliophyta
Class: Liliopsida
Subclass: Commelinidae
Order: Cyperales
Family: Poaceae
Genus: *Zea* L.

It is a wild grass in the Poaceae family that includes the species *Zea mays*.

8.2.1 CLASSIFICATIONS

a. Wilkes (1967): Based on ethnobotany, geography, cytology, and several morphological aspects—four races of teosinte for Mexico (Nobogame, Mesa Central, Chalco and Balsas) and two for Guatemala (Guatemala and Huehuetenango)
b. Doebley (1990): Based on hierarchical system of classification for *Zea* and morphological, ecological, and molecular features of the taxa. Divided *Zea* into two sections—(1) section *Luxuriantes* that includes *Zea perennis* (Hitch. Reeves and Mangelsdorf), *Zea diploperennis* (Iltis, Doebley, and Guzman), *Zea luxurians* (Durieu and Ascherson) Bird, and *Zea nicaraguensis* (Iltis and Benz).

    (2) Section *Zea* that includes *Zea mays* L. is divided into: *Zea mays* subsp. *Mexicana* (Schrader) Iltis for races Chalco, Mesa Central, and Nobogame; *Zea mays* subsp. *parviglumis* Iltis and Doebley that include races Balsas; *Zea mays* subsp. *huehuetenangensis* (Iltis and Doebley) Doebley for the teosinte race Huehuetenango, and *Zea mays* L. subsp. *mays* for cultivated maize.

c. Based on evidence from multiple independent sources, reported three new taxa—three new taxa from Mexico within section *Luxuriantes* from Nayarit, Michoacan and Oaxaca, Mexico.

TABLE 8.1 Some Species of *Zea* Their Habit, Distribution, and Ploidy.

Name	**Life cycle**	**Occurs in**	**2n = 2×**	**Ploidy**
Zea mays subsp. *Parviglumis*	Annual	Oaxaca state, Mexico	20	Diploid
Zea mays subsp. *huehuetenangensis*	Annual	Border area of Chiapas, Mexico and Huehuetenango, Guatemala	20	Diploid
Zea mays subsp. *Mexicana*	Annual	Central and Northern Mexico	20	Diploid
Zea mays subsp. *luxurians*	Annual	Guatemala, Honduras and Nicaragua	20	Diploid
Zea mays subsp. *diploperennis*	Perennial	Jalisco and Mexico	20	Diploid
Zea perennis	Perennial	Jalisco and Mexico	40	Tetraploid
Zea nicaraguensis	Annual		40	Tetraploid

8.3 INTRODUCTION

Teosinte name is commonly called for a group of four annual and perennial species of the genus *Zea* (native of Mexico and Central America) (Doebley, 1990). The name, teosinte, is of Nahuát Indian origin language and it has been interpreted to mean "grain of the gods." Teosinte, the closest relative of maize is used extensively for green fodder but stover, silage, and hay is also used for livestock. Teosinte resembles with maize phenotypically but differs in cob architecture, tillering, and branching habit. The cultivated maize is the results of domestication by humans over thousands of years from a series of hybridizations between the wild ancestors such as teosinte and trisacum. It is mostly adapted well to humid tropics and subtropics (Matsuoka et al., 2002).

Teosinte (*Euchlaena Mexicana* Schräd) is used as feed for animals due to its vigorous growth, high yield, and tolerance to biotic and abiotic stresses. In comparison to fodder crops grown in summer season such as pearl millet, sorghum, maize, etc., teosinte stays green for longer period of time, thus ensuring green herbage for longer duration (ICAR, 2011). Teosinte is grown mostly in subtropical areas of Central America, Mexico, Guatemala, and Nicaragua and it was supposed to be introduced in India in 1893, at Poona (Relwani et al., 1968). It occupied an area of about 10,000 ha in India (ICAR, 2011). The crop is well adapted to high humidity, rainfall, and can be grown with equal success in acidic and waterlogged soils. The quality of fodder is comparable to maize and possesses high percentage of carbohydrate. As a fodder crop, it can be cultivated in any type of fodder production system due to its maximum biomass production ability.

8.4 PLANT DESCRIPTION

Teosinte is an annual forage plant closely related botanically to corn, having tall and broader leaves. It produces tillers profusely; dark green narrow leaves which remain green for a longer period. The inflorescence is much like maize in which tassel is borne at the apex but no true ear is formed. Seeds are produced on slender spikes in four or five leaf axils, near the center of the plant. The height of the plant may be up to 9–15 ft, and bearing an abundance of leaves and tender stems. Interestingly, the difference between teosinte and maize is about five genes and they are compatible to produce teosinte × maize hybrids (Beadle, 1930). Later, DNA study from teosinte–maize offspring also confirmed the concept of five regions of the genome (which could be single genes or groups of genes), thus seemed to be controlling the most significant differences between teosinte and maize.

8.5 ORIGIN AND DISTRIBUTION

Teosinte grows in the valleys of Southwestern Mexico. In these regions, it grows commonly as a wild plant along streams and on hillsides. It is most common in the Balsas River drainage of Southwest Mexico and hence, is also known as Balsas teosinte. The distribution of teosinte is restricted to the western escarpment of tropical and subtropical areas of Mexico, Guatemala, Honduras, and Nicaragua, with isolated population limited areas varying in size from less than 1 ha to several square kilometers.

8.6 PLANT CHARACTERISTICS

Teosinte plants are taller and broader leaved than most grasses. Their general growth form is like that of maize, but teosinte produces lateral branches. Teosinte plants have main stalks that typically bear lateral branches at most of the nodes. The lateral branches and the main stalks are both having a series of nodes and internodes with a leaf attached at each node. Teosinte plants are branched, having several ears in two inter-leaved rows fashion where only 6–12 kernels were found, each sealed tightly in a stony casing, collectively known as a fruitcase. The leaves along the lateral branches are fully formed and composed of two parts, a sheath that clasps around the stem and a blade that extends away from the plant. The leaves on the branches are arranged in an alternate phyllotaxy. Both the main stem and the primary lateral branches of the teosinte plant possess male inflorescences (tassels), and female inflorescences (ears) are borne on secondary branches. Each of these female inflorescences is surrounded by a single, bladeless leaf, or husk. At maturity, the teosinte ear disarticulates such that the individual fruitcases become the dispersal units. The ears occur in clusters of one to five (or more) at each node along the branch. As compared to maize, teosinte has some peculiar characteristics:

TABLE 8.2 Difference Between Maize and Teosinte.

Characteristics	Maize	Teosinte
Plant architecture	Single stalk	Several stalk
Kernels number/ear	>500	6–12
Glume	Soft	Hard
Seed color	White, yellow, orange, black, red, etc.	Dark color, grays, variegated white, black, brown

Teosinte plants are generally tolerant to various biotic and abiotic stresses that occur in maize. *Zea perennis* and *Zea diploperennis* have resistance to viral diseases to which all other *Zea* are susceptible (Nault, 1983). Similarly, *Zea diploperennis* sources of resistance to *Striga* spp., menacing root parasites of significant importance in much of Africa and parts of Asia (Rich and Ejeta, 2008); *Zea luxurians* and *Zea nicarasguensis* have flood-resistance traits, such as the capacity to form root aerenchyma (Mano and Omori, 2007; Mano et al., 2009).

8.7 CLIMATIC REQUIREMENT

Teosinte can be grown up to an altitude of 600–2000 m above mean sea level depending on the species. The crop is grown in climates ranging from temperate to tropic. It can grow well up to 16–22°C temperature depending upon the species. However, the species *Z. nicaraguensis* has adapted up to 27°C. An annual rainfall of 100–200 cm is sufficient to grow the crop (Matthew et al., 2012).

8.8 SOIL AND ITS PREPARATION

The suitable soil type is well-drained sandy loam to loam soils with a pH range of 5.8–7.0. For sole cropping, the field should be prepared by three to four plowing to make weed-free field or two plowings with cultivator and one plowing with harrow can also be done. Teosinte is tolerant to water logging for some interval and is also suitable for acidic soil.

8.9 VARIETIES

Efforts on breeding for teosinte varieties are very less; hence a limited number of varieties are popular in different countries. Some of the important varieties/lines of India are given below.

TABLE 8.3 Varieties and Yield Range.

Varieties/lines	Yield range (t/ha)
Improved teosinte	40–50
TL 1 and TL 16	35–45
Sirsa, TL 2, Rhuri, Maizente-1, Maizente-2, Sirsa improved, TL 1	38–43

8.10 SOWING TIME

This crop is grown in various parts of the world. Under irrigated condition, the crops can be sown in winter and summer seasons. But, for rainfed condition, monsoon is suitable. In summer, sowing is carried out from March to mid-April, whereas in monsoon, June–July is good for production.

8.11 SEED TREATMENT AND SEED INOCULATION

Teosinte seeds have hard seed coat and show dormancy for some period (8–20). In Mexican teosinte populations, the dormancy is more than 90% due to which seed germination is may be up to 50–60% only. The species *Z. parviglumis* is deeply dormant while, *Z. luxurians* is weakly dormant (López et al., 2011). The germination percentage may be improved up to 30% by soaking the seeds in water for 24 h. A combination of seed treatment with water soaking (24 h) and biofertilizers (*Azospirillum* spp., *Cyanobacteria*, *Pseudomonas* spp.) provides up to additional 10% more germination. To protect the seed from diseases agrosan GN or Thiram @ 3.0 g/kg seed may be used.

8.12 SEED RATE AND SOWING METHOD

Direct sowing is done at row spacing of 25–30 cm using a seed rate of 35–40 kg/ha followed by heavy planking. Teosinte can be grown under no tillage condition to obtain the same green fodder yield as after conventional or zero till.

8.13 CROPPING PATTERN

The following cropping pattern is suitable to teosinte for higher productivity.
Teosinte + cowpea–berseem + mustard–sorghum + cowpea and
Teosinte + cowpea–berseem + mustard–maize + cowpea.

8.14 NUTRIENT MANAGEMENT (MANURES AND FERTILIZERS)

This crop is very responsive to fertilizers and manures present in soil. About 10–15 t/ha farmyard manure can be used before preparing the field. For fodder purpose, the major nutrient such as, nitrogen (N) is used @ 60 kg/ha, out of which basal application of 30 kg/ha is recommended and another 30 kg/ha is applied after 30 days after sowing (DAS). About 30 kg/ha P_2O_5 as basal dose is required for quality production of teosinte. Some species are multicut where the dose of nitrogen will increase to twice (120–150 kg N/ha) of no-cut system. One-third of it should be applied as basal dose, one-sixth after first irrigation, and one sixth after first, second, and third cut. The green fodder and dry matter yield will increase if the quantity of N is increased.

8.15 WATER MANAGEMENT

Depending upon the soil and weather, it needs irrigation at an interval of 8–10 days. Further, in case of long dry spell, rainy-season crop may also need 1–2 irrigations and in summer-sown crop, 7–8 irrigations are required.

8.16 WEED MANAGEMENT

The major weeds population prevalent in teosinte cultivation is *Commelina benghalensis*, *Digera arvensis*, *Portulaca quadrifida* and *Amaranthus* spp., *Eleusine*, *Aegyptiaca*, *Dactyloctenium aegyptium*, *Echinochloa colonum* and Cyperus rotundus. At 20–30 DAS hoeing with weeder controls the weeds effectively. For chemical management (1) preemergence application of simazine @ 1.50 kg a.i./ha in combination with mechanical weeding is proved to be most effective in controlling weeds and increasing the yield of fodders and (2) postemergence application of amine salts of 2,4-D and MCPA is effective. Mechanical weeding and intercropping with cowpea also provides good cultural control over weed to crop (Panday et al., 1969).

8.17 DISEASE AND INSECT-PEST MANAGEMENT

No any major disease and pest occurs in teosinte. Rather, this crop has various resistance genes for different biotic and abiotic stresses. But under severe condition, maize borer attacks the crop at the early stages of its growth which can be controlled by uprooting the plant and destroying them. Chemical controls include carbaryl 50WP at 250–300 g/ha against this pest. Before fodder is fed to animals, a gap of 21 days should be maintained after the spraying of the crop with chemicals. The disease and pest in this crop is not apparent but, can be seen similar as it also appears in maize.

8.18 HARVESTING AND YIELD

For production of succulent and nutritious fodder, the first cut should be taken 45–60 DAS and subsequently 30–45 days after previous harvest. It is best cut when 4 or 5 ft high, as it becomes less palatable if allowed maturing much beyond this. In single cut management, it should be harvested about 1 week before tasseling while, in multicut management, first cutting should be taken at 1.0–1.25-m height. For seed production purpose, the plants are sometimes cut once or twice only. The yields of forage are enormous, placing teosinte at the head of the grasses in the yields per acre. The average green fodder yield is about 500–700 quintals/ha under good management practices.

8.19 SEED PRODUCTION

Seed crop should be sown in the last week of June to the first week of July in rows 45-cm apart with a seed rate of 8–10 kg per acre. The crop should be harvested when three-fourth seeds mature. If harvesting is delayed, the mature seeds shatter which are loosely held. The sun-dried crop is thrashed by simple beating with sticks or by running a tractor over it. The white seeds are immature, which should be separated before storage. The average seed yield is about 5 q per acre. Harvests of 18 to 30 t per acre are not uncommon. It can produce 0.5–0.8 t/ha seed. The seeds of teosinte are different from that of the related species like maize. It is a highly cross-pollinated crop, thus requires:

- Rouging of off-type plants during critical stages of the crop (knee height, branch initiation time, flowering time, and maturity).
- Isolation distance should be maintained up to 400–500 m.
- Inspection at critical stages, and compulsory inspection at flower initiation to fertilization, would be effective for quality seed production.
- Bagging of large muslin cloth (that covers the whole inflorescence) or hand pollination (selfing and seibing) may be preferred for inbred seed production.

8.20 NUTRITIVE VALUE

Fodder quality such as crude protein, crude fiber, ether extract, and calcium content in teosinte can be improved by N application from 90 to 150 kg N/ha (Abichandani et al., 1970; Singh et al., 1988). The kernels are naked without adequate protection from predation and are easily digested by any animal that consumes them. Since the kernels are firmly attached to the cob and the ear does not disarticulate, a maize ear left on the plant will eventually fall to ground with its full set of kernels. It is comparatively less nutritious and palatable than maize but, due to its profuse tillering capacity it gives very good fodder yield.

8.21 UTILIZATION

The grain of this crop is not edible for human consumption. Commercially this crop is utilized for green fodder, dry fodder, and production of biofuels just like maize. For research purpose, we can utilize these wild species for transfer of various resistant genes to cultivated species

TABLE 8.4 Quality and Yield Range in Teosinte Fodder.

Parameters	Percentage (%)	Parameters	Percentage (%)
Dry matter	22.0–24.0	Total carbohydrate	82.0–87.0
Crude protein	6.0–7.0	Neutral detergent fiber	60.0–63.0
Ether extract	1.4–2.0	Neutral detergent fiber	30.0–33.0
Ash	6.0–10.0	Hemicellulose	29.0–30.5
Crude fiber	28.0–30.0	ME	2.4 M Cal/kg
Nitrogen free extract	50.0–55.0	Green fodder yield	30.0–50.0 t/ha
Organic matter	90.0–94.0	Dry matter yield	4.0–6.0 t/ha
In vitro dry matter digestibility	50.0–55.0		

ME, metabolizable energy; M Cal/kg, mega calories per kilogram.

of maize. Indian Grass and Fodder Research Institute (IGFRI), Jhansi has developed various line of maize, derived from teosinte × maize or maize × teosinte crosses for development of fodder traits.

KEYWORDS

- **teosinte**
- **agronomic management**
- **dry matter yield**
- **green fodder production**
- **nutritive value**

REFERENCES

Abichandani, C. T.; Manikar, N. D.; Mishra, M. N.; Karnail, G. T.; Gill, A. S.; Maurya, R. K.; Chandra, G. *Effect of Top Dressing of N on Fodder Yield of Teosinte*; Annual Report for I.G.F.R.I.: Jhansi, 1970; p 50.

Beadle, G. W. *Genetical and Cytological Studies of Mendelian Asynapsis in Zea mays*; Cornell University Agricultural Experiment Station Memoir; Cornell University: Ithaca, New York, 1930; Vol. 129, 1–22.

Doebley, J. F. Molecular Evidence and the Evolution of Maize. *Econ. Bot.* **1990,** *44* (3), 6–27.

ICAR. *Handbook of Agriculture*; Indian Council of Agricultural Research (ICAR): New Delhi, 2011; pp 1353–1417.

López, A. N. A.; Jesús Sánchez González, J.; Ruíz Corral, J. A.; De La Cruz Larios, L.; Santacruz-Ruvalcaba, F.; Sánchez Hernández, C. V.; Holland, J. B. Seed Dormancy in Mexican Teosinte. *Crop Sci.* **2011,** *51*, 2056–2066.

Mano, Y.; Omori, F. Breeding for Flooding Tolerant Maize Using "Teosinte" as a Germplasm Resource. *Plant Root* **2007,** *1*, 17–21.

Mano, Y.; Omori, C.; Loaisiga, H; Mck, B. R. QTL Mapping of Above-ground Adventitious Roots During Flooding in Maize × Teosinte "*Zea nicaraguensis*" Backcross Population. *Plant Root* **2009,** *3*, 3–9.

Matsuoka, Y.; Vigouroux, Y.; Goodman, M. M.; Sanchez, J.; Buckler, E. S.; Doebley, J. F. A Single Domestication for Maize Shown by Multilocus Microsatellite Genotyping. *Proc. Natl. Acad. Sci. USA* **2002,** *99*, 6080–6084.

Matthew, B. H.; Bilinski, P.; Pyhäjärvi, T.; Ibarra, J. R. Teosinte as a Model System for Population and Ecological Genomics. *Trends Genet.* **2012,** *28* (12), 606–615.

Nault, L. In *Origins of Leafhopper Vectors of Maize Pathogens in Mesoamerica*, Proceedings of International Maize Virus Disease Colloquium and Workshop, Wooster,

Ohio, August 2–6, 1982; Gordon, D. T., Knoke, J. K., Nault, L. R., Ritter, R. M. Eds.; The Ohio State University, Ohio Agricultural Research and Development Center: Wooster, Ohio, USA, 1983; pp 75–82.

Panday, R. K.; Singh, R. P.; Singh, M. Weed Control in the Fodder Crops of Teosinte and Maize. *Indian J. Weed Sci.* **1969,** 1 (3–4), 95–102.

Relwani, L. L., Bagga, R. K.; Walli, T. K. *Annual Report*; NDRI: Karnal, 1968.

Rich, P. J.; Ejeta, G. Towards Effective Resistance to *Striga* in African Maize. *Plant Signal. Behav.* **2008,** *3*, 618–621.

Singh, V.; Sharma, S. P.; Verma, S. S.; Joshi, P. Effect of Nitrogen Fertilization and Cutting Management on the Yield and Quality of Herbage of Teosinte. (*Euchlaena Mexicana* Schrad.). *Trop. Agricult.* **1988,** *65* (3), 194–196.

Wilkes, H. G. *Teosinte: The Closest Relative of Maize*; Bussey Institute, Harvard University: Cambridge, MA, 1967.

CHAPTER 9

SUDAN GRASS (SUDAN SORGHUM)

ANUPAM MUKHERJEE[1*] and MD. HEDAYETULLAH[2]

[1]*Sasya Shyamala Krishi Vigyan Kendra, Ramakrishna Mission Vivekananda University, Arapanch 700150, Sonarpur, Kolkata, India*

[2]*Department of Agronomy, Faculty of Agriculture, Bidhan Chandra Krishi Viswavidyalaya, Mohanpur 741252, Nadia, West Bengal, India*

[*]*Corresponding author. E-mail: anupammukhe@gmail.com*

ABSTRACT

Sudan grass is an annual forage grass native to areas throughout Africa and Southern Asia. The Sudan grass originates from Sudan and Southern Egypt. Sudan grass has low tolerance for cold. During cold periods, if plants are not killed, they remain dormant and resume growth only when conditions are favorable. Sudan grass is highly palatable to all kinds of livestock and dairy animals. The average crude protein content is about 12%; crude fiber is less than 30%, and NFE is about 40–45%. Sudan grass is grown for pasture, grazing, green chop silage, hay, or seed. The crop is used as pasture for animals such as cows, sheep, and hogs, and as a range plant for poultry, especially turkey. The average green fodder yield is about 400–500 quintals/ha and with irrigation it often reaches 600 quintals/ha from single cut management. The average seed yield varies between 4 and 5 quintals/ha from October–November harvests and between 6 and 8 quintals/ha from April–May harvests.

9.1 BOTANICAL NAME AND CLASSIFICATION

Botanical name: *Sorghum vulgare* var. sudanense
Family: Poaceae (Gramineae)
Subfamily: Panicoideae
Tribe: Andropogoneae
Genus: *Sorghum*
Species: *vulgare*
Variety: sudanense

9.2 COMMON NAME

Sudan grass, Sudan sorghum.

9.3 INTRODUCTION

Sudan Grass is an annual forage grass native to areas throughout Africa and Southern Asia. The Sudan grass is a heavy feeder, as would be expected from the large tonnage secured in a season. It contains about 10% protein, 30% crude fiber, and 9.7% ash with calcium and phosphorus in proper balance (Aguiar et al., 2006). This is a valuable forage plant which grows up to 10 ft. The thick and erect stems are usually arising in groups from a single clump. The leaves are long and narrow and arranged at the end of the stems on loose-bending branches. The flowering parts are in the form of a loose panicle at the end of the stem. The hulls are awned and often purple in color. Sudan grass is well adapted to the drier areas of the country for its tolerance to long hot and dry periods of weather.

9.4 ORIGIN AND DISTRIBUTION

The Sudan grass originates from Sudan and Southern Egypt. It was first introduced to the United States by C. V. Piper in 1909. Sudan grass is widely adapted as a cultivated grassy fodder crop. In India, it was introduced first to northeastern state; Assam in 1920. Sudan grass is popularly grown in South Africa, Southern Europe, and South America.

9.5 CLIMATIC REQUIREMENTS

Sudan grass prefers a worm climate of relatively low humidity. Sudan grass has low tolerance for cold. During cold periods, if plants are not killed, they remain dormant and resume growth only when conditions are favorable. Plants cannot withstand frosts during the growing season. The crop grows poorly at higher altitudes due to cool or cold weather conditions and untimely frosts. In periods of drought, the plant becomes dormant and resumes growth when moisture conditions improve. However, extended periods of drought can cause wilting

9.6 SOIL AND ITS PREPARATION

The crop grows successfully on almost every type of soil, but it does best on loams. Coarse, porous, sandy, or gravelly soils are generally not good for production. Usually, soils that are good for growing sorghum are also good for Sudan grass production. Even though the crop is not particularly sensitive to acidic soils, when soil pH falls below 5.5, lime is recommended for soil application.

9.7 SEEDBED PREPARATION

The seeds of Sudan grass are rather small so a well-prepared seedbed is necessary to insure a good germination and better stand. If the field is free from weeds, the seedbed may be prepared with a disc and a harrow. A well-prepared, firm, moist seedbed is better but acceptable stands can grow in grass sods, stubble, or where you use no till.

9.8 VARIETIES

9.8.1 MEETHI SUDAN

This variety is an outcome of the cross involving sweet jowar JS 263. It is a tall plant having a height of 250–300 cm, vigorous, thin stemmed, sweet, and profuse tillering. It is leafy with green midrib and has high leaf:stem ratio. Fifty percent of flowering is attained in 50–55 days and maturity in 90–95 days. Panicles are very loose, open glumes awned, and straw colored. The variety is moderately resistant to red leaf spot. The average green fodder yield is 77.5 t/ha. The variety was released in 1976.

9.8.2 PUNJAB SUDEX CHARI-1 (LY-250)

This variety is a sorghum–sudex hybrid derived from sorghum (2077 A) × Sudan grass (SGL-87) hybridization followed by selection. It is a multicut variety providing three to four cuts. Stem is thin and palatable. The variety is highly resistant to anthracnose. It produces 95 t/ha of green fodder and 25 t/ha dry matter. The variety was released in 1995.

9.9 SOWING

9.9.1 TIME OF SOWING

Purchase seed that is adapted to your locality. In Eastern region, Sudan grass can be planted between first week of May and first week of June. It is important to wait until soils are warm enough before seeding. The cool soil results the slow growth and gives weed seeds advantage over the crop. Sow Sudan grass when soil temperatures are at least 15°C. The best time for seeding Sudan grass, however, will necessarily vary with the location and to some extent with the type and characteristics of the soil. Late planting is not recommended, especially if Sudan grass is grown for seed. Very hot summers or early fall rains reduce yields.

9.9.2 SEEDING METHOD

Sudan grass may be seeded either in cultivated rows or in close drills for hay purposes. In the drier sections, better results will be obtained by seeding in cultivated rows having space of 18 in. or more apart. This method provides the individual plants with a greater area from which to receive sufficient moisture. Under high rainfall or irrigated conditions, use a row spacing of 6 in. or more. If the crop is being grazed by animals, use at least 20-in. spacing between rows to allow animals to move in the field without trampling the crop. Depending on available moisture and soil type, planting depths can range from 1 to 1.5 in. Cover seed with soil to enhance germination. Growers usually drill or broadcast Sudan grass at a rate of 11–15 kg of pure and live seeds per acre. Generally, lower seeding rates are better for dry areas. Higher rates are better for irrigated plantings or where rainfall is adequate. Research has shown that total forage yield is about the same regardless of the row spacing, because both Sudan grass and its hybrids tiller extensively.

9.9.3 SPACING

The row to row distance varies from 25 to 40 cm in moderate rainfall areas and 50 cm in dry areas.

9.9.4 SEED RATE

Seed rate varies from 15 to 75 kg/ha. The lower seed rates are maintained in for dry condition and higher seed rates are maintained in humid conditions. The normal seed rate is 25–35 kg/ha.

9.10 SEED TREATMENTS AND INOCULATION

The seed should always be purchased from a reliable source. If seed is not already treated, seed may be treated with fungicides like Thiram or agrosan GN or a combination of carboxin and Thiram @ 3 g/kg of seed. For nitrogen fixation, seed may be inoculated with azotobacter @ 20 g / kg seed.

9.11 CROP SEQUENCE AND CROP MIXTURE

Sudan crop belongs to the family Poaceae (Gramineae); therefore, it is sown as a sole crop followed by leguminous fodder crops. In crop mixture, Sudan grasses are grown in mixed cropping or intercropping systems with leguminous fodder crops.

9.11.1 CROP SEQUENCE

i. Sweet Sudan grass–berseem + mustard
ii. Sweet Sudan grass–berseem + oats
iii. Sweet Sudan grass–maize + cowpea
iv. Sweet Sudan grass–jower + cowpea

9.11.2 CROP MIXTURE

i. Sweet Sudan grass+ cowpea
ii. Sweet Sudan grass + guar

9.12 MANURE AND FERTILIZER APPLICATION

Unlike legumes, it is dependent solely on the soil for its source of nitrogen. During land preparation 10–20 t farmyard manure is recommended for 1 ha of land for higher green fodder yield. Use soil tests to determine the crop's nutrient needs. Fertilizer requirements are usually similar to those of other annual grass crops. Nitrogen (N) is the most limiting nutrient in Sudan grass production. Phosphorous moves slowly in the soil so apply it before planting or band it at seeding. Adjust the amount of fertilizer you need based on the previous crop. Adjust rate and time of application of fertilizer to moisture supply and forage needs for optimum production. Split applications are important in order to reduce the possibility of nitrate poisoning in animals. Divide applications equally (depending on the number of harvests) and apply after each grazing or harvesting operation. The recommended dose of nitrogen, phosphorus, and potash in kg acre^{-1} are 100–120:25–30:30–40, respectively out of which 45–50 kg of nitrogen are applied at the time of land preparation. After each cut, 20 kg acre^{-1} more nitrogen should be applied to encourage growth.

9.13 WATER MANAGEMENT

When Sudan grass is used as a forage crop it will be benefited under irrigation. Again, in sections where the precipitation is low and the supply of irrigation water is limited, much larger yields of forage will often be obtained by applying the small amount of water. When Sudan grass is planted early in the spring, it will be necessary to irrigate the ground before seeding, as the natural moisture will usually be sufficient to germinate the seed. Ordinarily three or four irrigations throughout the growing season will be sufficient under most conditions. The first irrigation should be given when the plant reaches a height of 18–25 in. or just before heading. The later irrigations may be applied after each cutting or as soon as the new growth has started (Madson and Kennedy, 1917).

9.14 WEED CONTROL

In initial stage, intercultural operations are necessary to keep down weeds while the grass is getting established. According to need, two to three weeding are required to control the weeds. Weeding may be done either with hand hoe or wheel hoe. Plant weed-free seed because weed problems most often come from the seed source. Buy seed from reputable dealers. Also, check the seed tag to determine percent seed purity. For better weed management use integrated approach of weed management which contains both cultural and chemical methods of weed control. Glyphosate at 2.0–2.5 kg a.i. ha^{-1} before planting or at the time of final land preparation successfully controls the broad spectrum of weeds. Atrazin 1 kg a.i. ha^{-1} and pendimethalin 1 kg a.i. ha^{-1} as preemergence also helps in controlling weeds. But before herbicide application, please ensure the dose and time from any recognized source.

9.15 INSECT-PEST AND DISEASE MANAGEMENT

9.15.1 INSECT-PEST

Cutworms frequently injure Sudan grass. They are especially injurious to seedlings, cutting them off above or below the soil surface. They are most active at night. A full-grown cutworm is about 1–2-in. long, smooth, and either striped green, brown, or gray or mottled in color. Grasshoppers also harm Sudan grass. Eastern region is especially susceptible to migratory grasshopper attacks. Chinch bugs and green bugs attack Sudan grass in other states. You can spray with approved insecticides if these insects reach economic thresholds. You can also use the pheromone trap and light trap for controlling these insect-pest attacks.

Pheasants and quail dig seeds out of the ground, reducing stand establishment. Crows and pheasants feed on seedlings. Blackbirds and sparrows feed at seed ripening stage. Birds are most destructive when the crop is grown for seed. Birds are most difficult to control. There is no single effective method. To minimize damage, do not plant fields near timbered or scrub brush areas. Ground squirrels, woodchucks, mice, jackrabbits, and gophers can cause considerable harm. Field mice usually eat planted seeds. Moles and pocket gophers do a lot of underground damage. You can use film strips, plastic bags, and tin boxes for producing sound for controlling birds, squirrels, moles, and gophers.

9.15.2 DISEASES

The major diseases affecting Sudan grass plants are kernel smut, leaf blights, and downy mildew. The disease kernel smut is first noticeable after the Sudan grass has begun to head. Examination at this time shows that many of the seeds are replaced by smut balls or "false kernels." Leaf blights cause elongated, straw-colored lesions with reddish margins on leaves. The best control is to use resistant varieties, if they are available. Rotations with other crops break the disease cycle. Plants infected with downy mildew have deformed yellowish and reddish leaves. Once again, use of resistant varieties is the best control method. Also, remove infected debris from the fields. Treat seeds to help control kernel and head smut. Provide the crop optimum growing conditions to help minimize the outbreak of diseases.

9.16 HARVESTING

9.16.1 PASTURE

Graze the crop when plants reach a height of at least 18 in. 6–8 weeks after seeding to reduce the risk of nitrate poisoning. Stock the pasture enough to graze it down before heading begins. Graze the pasture rapidly to 6–8-in. stubble. Then, discontinue grazing until growth again reaches at least 18-in. high, and graze again. It usually takes 3–4 weeks for sufficient regrowth. Consider using at least a 20-in. row spacing to allow cattle room to move in the field without trampling the crop. Short rotational grazing systems work well. Subdivide fields into three or more pastures, and graze down each pasture in 7–10 days.

Stagger the planting dates for each pasture about 10 days apart so that grazing can begin on each pasture when the Sudan grass is at the right height. Animals trample and feed selectively when growth reaches 40-in. high or higher. If you cannot set cattle to graze the field at the proper time, harvest the crop for hay or silage.

9.16.2 GREEN CHOP

This is a high-protein forage. Growers usually feed it to dairy cattle and other high-producing livestock. Harvest plants when they are 18–24-in. tall. Harvest before heading, as dry matter intake and digestibility are reduced after this stage. Feed harvested forage to animals immediately.

9.16.3 HAY

In order to obtain the good quality of hay, Sudan grass should be cut when in full bloom. If harvesting is delayed until the seed is ripe, the hay will be a little coarse and a little less palatable than if cut earlier. The most common way of harvesting the hay crop is with a mower. During the bright, warm days of the summer, grass cut one morning will usually be ready to rake into windrows or bunches the next day. To produce a bright hay of high quality, it should be put into cocks and allowed to cure for 3 or 4 days before being put into the mow or stack. The yields of hay which may be obtained with Sudan grass will vary with the soil condition, the moisture level, and the other factors which enhance or retard plant growth. During the growing period of Sudan grass, the yields of hay have varied from 5 to 9 t per acre. Yields are greatest when the heads reach the soft dough stage. However, thick stems at this stage make drying difficult, so it is best to cut before heads emerge. Sudan grass hay fodder may yielded from single cut is 16 t per hectare (Sowinski and Szydelko, 2011).

9.16.4 SEED

The yield of seed will be variable according to the character of the soil, the supply of moisture, and the climatic condition during the growth period. The average seed yield of Sudan grass varies from 200 to 250 kg per acre in dry condition and it ranges from 300 to 325 kg per acre at irrigated conditions. Seed should be uniformly mature before harvest. Sudan grass seed heads normally do not shatter, so harvesting the crop is comparatively easy.

9.17 GREEN FODDER YIELD

The average green fodder yield is about 400–500 quintals/ha and with irrigation it often reaches 600 quintals/ha from single cut management. Application with a high dose of N fertilizers and irrigation, the green herbage yield may be as high as 1200 quintals/ha in three to four cuts. The average hay yields may be 150–200 quintals/ha.

9.18 SEED PRODUCTION

Sudan grass is a photosensitive fodder crop to a certain extent. For seed production, seeds are sown in wider spacing to establish a healthy plant. In Punjab, India the ratoon crop is usually left for seed production. In India, the July-sown crop produces seed by October or November. But seeds are best collected from the ratoon crop in the month of April and May. The average seed yield varies between 4 and 5 quintals/ha from October to November harvests and between 6 and 8 quintals/ha from April to May harvests.

9.19 NUTRITIVE VALUE

Sudan grass is highly palatable to all kinds of livestock and dairy animals. The herbage is rich in crude protein. The herbage contents of crude fiber (CF) are about 8.5%, ether extract (EE) 0.6%, and nitrogen free extract (NFE) 12.8% where the digestibility of crude protein (CP) is about 72%, CF 76%, EE 72%, and NFE about 69%. But in hay, the average CP content is about 12%; of CF are less than 30% and NFE about 40–45%. Estimated nutrient composition of Sudan grass is given in Table 9.1.

TABLE 9.1 Estimated Nutrient Composition of Sudan Grass (% Is Dry Matter Basis).[a]

Forage	DM	CP	ADF	NDF	TDN	Ca	P
Fresh and early vegetative	18	16.8	29	55	70	0.43	0.41
Fresh and mid bloom	23	8.8	40	65	63	0.43	0.35
Hay and full bloom	91	8.0	42	68	57	0.55	0.30
Silage	28	10.8	42	68	56	0.46	0.21

ADF, acid detergent fiber; Ca, calcium; CP, crude protein; DM, dry matter; NDF, neutral detergent fiber; P, phosphorus; TDN, total digestible nutrients.

[a]Adapted from National Research Council, 1989.

9.20 UTILIZATION

Sudan grass is grown for pasture, grazing, green chop silage, hay, or seed. The crop is used as pasture for animals such as cows, sheep, and hogs, and as a range plant for poultry, especially turkey. As a pasture crop, Sudan grass has a higher carrying capacity than other annual grasses or legumes, especially in regions with hot and dry summers. Growers can easily make five cuttings in a year where the growing season is long. They also can make it as an emergency crop when the other forage crops have failed, which fills an important need in many regions.

9.21 POTENTIAL HAZARDS (TOXICITIES)

9.21.1 HYDROGEN CYANIDE OR PRUSSIC ACID POISONING

Sudan grass is sometimes considered dangerous to grazing livestock because it may contain hydrocyanic acid which is a toxic substance. The acid content of the grass may also be decreased by limiting planting to fertile soils with large water-holding capacity. This poisoning is eliminated by delaying grazing until the plants reach a height of 91 cm (36 in.) or more, at which stage they are relatively free of the substance. Take the following precautions to avoid this problem.

9.21.2 NITRATE POISONING

Animals can be poisoned if they ingest forage containing high concentrations of nitrates. These compounds change in the digestive tract and are absorbed into the blood stream, where they interfere with oxygen transport. Symptoms are the labored breathing, muscle tremors, and staggering. Take the following precautions to avoid nitrate poisoning:

i. Apply N fertilizer in split application.
ii. Graze when plants are at least 18 in.
iii. Do not graze at night during frost.
iv. Delay feeding silage for 40–50 days after ensiling.

KEYWORDS

- **Sudan grass**
- **agronomic management**
- **fodder yield**
- **nutritive value**
- **seed production**

REFERENCES

Aguiar, E. M. de.; Lima, G. F. da C.; Santos, M. V. F. dos.; Carvalho, F. F. R. de.; Guim, A.; Medeiros, H. R. de.; Borges, A. Q. Yield and Chemical Composition of Chopped Tropical Grass Hays. *Rev. Bras. Zootec.* **2006,** *35* (6), 2226–2233.

Madson, B. A.; Kennedy, P. B. *Sudan Grass*; University of California Publications, University of California Press: Berkeley, Bulletin No. 277, 1917.

National Research Council. *Nutrient Requirements of Dairy Cattle,* 6th ed.; National Academy of Sciences: Washington, D.C., 1989.

Sowinski, J.; Szydelko, E. Growth Rate and Yields of a Sorghum–Sudangrass Hybrid Variety Grown on a Light and a Medium-heavy Soil as Affected by Cutting Management and Seeding Rate. *Polish J. Agron.* **2011,** *4,* 23–28.

CHAPTER 10

HYBRID NAPIER (NAPIER BAJRA HYBRID)

JYOTIRMAY KARFORMA*

Regional Research Station (Old Alluvial Zone), UBKV, Majhian, West Bengal, India

**E-mail: jkarforma@gmail.com*

ABSTRACT

Hybrid Napier is an interspecific hybrid between bajra and Napier grass and combines high quality and faster growth of bajra with deep root system and multicut habit of Napier grass. It is widely distributed in subtropical regions of Asia, Africa, Southern Europe, and America. In India, it is mainly cultivated in Punjab, Haryana, Uttar Pradesh, Bihar, Madhya Pradesh, Odisha, Gujarat, West Bengal, Assam, etc. The grass grows throughout the year in the tropics. It can grow on a variety of soils. Light showers alternated with bright sunshine are very congenial to the crop. It produces large number of tillers and numerous leaves. The leaves are large and green, the sheaths are softer, and the margins are less serrated and hence, the herbage is palatable. It is juicer and succulent at all stages of crop growth. It is less fibrous and more acceptable. CO-1, IGFRI-5, CO-3, CO-4, and APBN-1 are improved varieties of hybrid Napier. One rooted slip or stem cutting is planted at a depth of 3–5 cm on one side of the ridge at 60 cm × 60-cm or 70 cm × 50-cm spacing. In a well established crop, the first cut is ready in 60–65 days after planting and subsequent cuttings are taken approximately at an interval of 25–30 days. Annually at least 6–8 cuts are possible. Generally it produces green forage yield of 150–300 t ha^{-1} $year^{-1}$.

10.1 BOTANICAL NAME AND CLASSIFICATION

Pennisetum americanum × *P. purpureum* (bajra × Napier) It is an interspecific hybrid between bajra (*Pennisetum americanum L.*) and common Napier grass (*Pennisetum purpureum Schum*). This grass belongs to the family *Poaceae* (*Gramineae*), subfamily *Panicoideae,* and tribe *Paniceae.*

10.2 COMMON NAME

Napier–bajra hybrid is often also referred as bajra–Napier hybrids, N–B hybrids, King grass, elephant millet, and Cumbu Napier hybrid and are tall growing (2–6 m), erect, stout, and deep rooted perennial hybrid grasses.

10.3 INTRODUCTION

Low productivity of dairy animals could be accredited to the less availability of forage along with its poor quality. To maximize the milk production, it is essential to feed animals with quality green fodder. Consequently, high-yielding forage including number of varieties of hybrid Napier has been developed. To increase the profit in dairy, farming growing forage crops is a new concept for most of farmers to supply green fodder to their dairy animals on a regular basis. Among all fodder crops hybrid Napier produces highest green forage yield and it grows throughout the year in the tropics under irrigated situation. Hybrid Napier is an interspecific hybrid between bajra and Napier. This fodder crop is having high-quality palatable grasses with deep root system and multicut habit.

10.4 PLANT DESCRIPTIONS

Hybrid Napier grass generally grows to a height of about 1.5–2.0 m but it can be raised up to of 6.0-m height. It is roughly similar to sugarcane in habit of growth. The plant tiller freely and single clump may produce more than 50 tillers under favorable climatic and soil conditions. This grass is succulent, leafy, fine textured, palatable and fast growing, and drought resistant. The leaves are large and greener, the sheaths are soft, and the margins are less serrated. It is juicy, less fibrous, and succulent at all the stages of crop growth. It has high sugar content and leaf:stem ratio. It grows faster and produces numerous leaves, thereby more herbage. Hybrid Napier is a perennial grass which can be retained on fields for 2–3 years. It is one of the highest yielding perennial tropical fodder grasses and considered as cut and carry forage for stall feeding system.

10.5 ORIGIN AND DISTRIBUTION

An interspecific hybrid between Napier grass and bajra was first produced in South Africa and released under the name, Babala Napier hybrid or Bana grass. Hybridization works were started in India at Coimbatore (1953) and then at New Delhi (1961) resulting in the release of Cumbu Napier hybrid and Pusa giant Napier respectively. It is one of the highest yielding perennial tropical fodder grasses and considered as a cut and carry forage for stall feeder system. During 1989–2000, various forage breeding institutions in the world have developed vast number of bajra × Napier hybrids. Scientists of the Southern African Development Community (SADC), International Crop Research Institute for the Semi Arid Tropic (ICRISAT), Sorghum and Millet Improvement Program (SMIP) in Zimbabwe and Punjab Agricultural University and Tamil Nadu Agricultural University (TNAU) in India were the pioneers of developing vast number of bajra × Napier hybrids. Hybrid Napier var. CO-3 was developed by the scientists at TNAU at Coimbatore and released for commercial cultivation in 1997. Later Animal Husbandry Department, Government of Tamil Nadu has taken up activities to popularize this grass among the farmers. It is widely distributed in subtropical regions of Asia, Africa, Southern Europe, and America. It is the major forage crop in the tropics and subtropics of the world. It is well adopted in altitude ranged from sea level to 2000 m and latitude between 10° N and 20° S. In India, it is mainly cultivated in Punjab, Haryana, Uttar Pradesh, Bihar, Madhya Pradesh, Odisha, Gujarat, West Bengal, Assam, Andhra Pradesh, Kerala, and Tamil Nadu.

10.6 PLANT CHARACTERISTICS

The interspecific hybrid is a triploid with $2n = 3\times = 21$ chromosomes (seven chromosomes from pearl millet and 14 chromosomes from Napier grass). Triploid hybrid resulting from the interspecific cross is usually highly variable because of the heterozygosis of Napier grass. The hybrids generally show high heterosis for fodder yield and quality, and thus are high yielding and more acceptable than the Napier grass (Bogdan, 1977). However, these hybrids are sterile and need to be propagated vegetatively, which puts a major limitation on their easy distribution to farmers. Hybrid Napier has a perennial life cycle and a profuse root system, penetrating deep into the soil and an abundance of fibrous roots spreading into the top soil horizons. The rhizome (underground stem) is short and creeping and nodes develop fine roots and clumps. The plant forms clumps and grows upwards profusely with thick cane like stems. The culms are erect and tall, varying in height generally from 1.5 to 2 m but it may be up to 6 m in height. It has long tapering and stiff leaves. The upper leaf surface is covered with less persistent hairs and leaf sheaths are also hairy at minimum extent. The grass tolerates frequent defoliation, under good weather condition; it can be cut every 6–8 weeks giving up to 8 cuts in a year, depending on fertilizer applications, rainfall amount, and distribution.

10.7 CLIMATIC REQUIREMENTS

The grass grows throughout the year in the tropics. The optimum temperature for growth is usually 25–40°C and light showers alternated with bright sunshine are very congenial to the crop. The growth of the crop checks below 15°C. The grass is very susceptible to frost. Total water requirement of the grass is about 800–1000 mm. This grass grows best in high rainfall areas (in excess of 1500 mm per year).

10.8 SOIL AND ITS PREPARATION

Hybrid Napier can grow on a variety of soils. Light loams and sandy loam soils are preferred to heavy soils. It grows well on deep, retentive soils of moderate to fairly heavy texture and also grows on light textured soils. For better response to management, loamy soils with good drainage are preferred. The soil has to be wet at the root zone but should not be stagnated. The grass does not thrive well on water logged and flood prone lands. It prefers a deep fertile soil. Soils or lands, such as, eroded and rocky, water logged, salt affected, and very acidic should be avoided. Phenomenal yields are obtained from very deep fertile soil rich in organic matter and nutrient elements. It tolerates a pH range from 5 to 8.

As it is a fodder crop, it can be extended or cultivated in a fallow or barren land therefore, clearing of the bushes, removal of thorns, weeds, etc., are to be done at the beginning. It is a long duration crop, hence periodical tillage activities like other crops are not possible after the crop occupies the field. Generally two to three plowing followed by planking are required to obtain the fine tilth. Ridges are made across the slopes as far as possible at a spacing of 60 cm with a height of about 25 cm which enables irrigation uniform and easy.

10.9 VARIETIES

A number of improved varieties of hybrid Napier are developed at different State Agricultural Universities and Research Institutes. Some of them are described below:

CO-1: The variety was developed at TNAU, Coimbatore in the year 1984 and recommended for cultivation in south zone of the country. It is drought tolerant, high-yielding variety, nonlodging, profuse tillering, highly leafy, and high leaf:stem ratio. The average green fodder yield is 300 t ha^{-1}.

Hybrid Napier-3 (Swetika): This was developed at Indian Grassland and Fodder Research Institute (IGFRI), Jhansi and recommended for cultivation in north and central zone of the country. This is profuse tillering type, erect with narrow upright leaves with quick regeneration ability. It is tolerant to frost and low temperature and is suitable for low pH. It gives 70–80 t ha^{-1} green fodder and 18 t ha^{-1} dry fodders.

CO-2: The variety was developed at TNAU, Coimbatore and recommended for cultivation in south zone of the country. It provides 350 t ha^{-1} of green fodder.

CO-3: The variety was developed at TNAU, Coimbatore in the year 1997 and recommended for cultivation in south zone of the country under irrigated condition. The variety has high leaf:stem ratio making it highly palatable for animals. It gives 300 t ha^{-1} of green fodder.

PBN-83: The variety was developed at Punjab Agricultural University, Ludhiana and released for cultivation in Punjab.

IGFRI-5: It is suitable for cultivation in areas of submountain and low hill subtropical zone of Himachal Pradesh, which are below 800 m under rainfed and irrigated conditions and fertile as well as marginal lands. It is very vigorous and tall growing with thin tillers, erect growth habit, very leafy, glabrous nodes, and soft stems up to full growth.

NB-21: It was developed by Punjab Agricultural University, Ludhiana and released for cultivation throughout the India. It is fast growing variety with high tillering capacity. Stems are thin and nonhairy with long, smooth, and narrow leaves.

APBN-1: The variety was developed by Acharya N. G. Ranga Agricultural University, Hyderabad and recommended for cultivation in whole Gujarat. It has profuse tillering habit and high leaf:stem ratio. It gives 200 t ha^{-1} of fresh fodder.

Suguna: The variety was developed by Kerala Agricultural University, Vellayani in 2006. It is a semiperennial, multicut, and yield 260 t ha^{-1} per year green fodder. The variety is recommended for cultivation in southern districts of Kerala state.

Supriya: The variety was developed by Kerala Agricultural University, Vellayani in 2006. The variety is recommended for cultivation in southern districts of Kerala state and yields 270 t ha^{-1} green fodder.

10.10 SOWING TIME

As it is an interspecific hybrid and cannot be grown by using seeds. It is commercially cultivated by planting two budded stems or rooted slips. Planting can be done from February to August with the onset of monsoon for better establishment. Under irrigated conditions, planting can be done throughout the year.

10.11 SEED TREATMENT AND SEED INOCULATION

This interspecific hybrid is sterile and does not produce viable seeds although it could produce an inflorescence or spike. It is vegetatively propagated by rooted slips or stem cuttings and this helps to preserve its genetic purity. For quality planting materials, rooted slips and stem cutting are supposed to be in fresh status and with three to four active tillers and two nodes, respectively. Using the planting materials soon after removing them from mother plants is recommended. Pests and diseases are not the problem for hybrid Napier cultivation therefore; seed treatment or sett treatment is not required before planting.

10.12 SEED RATE AND SOWING METHOD

About 25,000–40,000 rooted slips or stem cuttings per hectare with spacing of 70 cm × 50 cm or 60 cm × 60 cm for sole cropping and 100 cm × 50 cm for intercropping are needed. Cuttings with two nodes from the middle portion of moderately matured stems (3–4 months old) are preferred.

One rooted slip or stem cutting is planted at a depth of 3–5 cm on one side of the ridge at above spacing in a single hole or spot. The cuttings are planted at a slanting position at one side of the ridges with one node buried in the soil. The underground node develops roots, while the upper node develops shoots only. The soil around the stem has to be pressed tightly for better root growth.

10.13 CROPPING PATTERN

It is the pattern of crops for a given piece of land or cropping pattern means the proportion of area under various crops at a point of time in a unit area or it indicates the yearly sequence and spatial arrangements of crops and follows in an area. Land resources being limited, emphasis has to be given for increasing production from unit area of land in a year. The important requisites in fodder production for intensive dairy farming are that fodder is required in uniform quantity throughout the year and the fodder crops in the rotation should be high yielding. The best rotation in the overlapping cropping system is berseem + Japan sarson–hybrid Napier + cowpea–hybrid Napier; (October–April)–(April–June)–(June–October). Hybrid Napier is introduced in the standing crop of berseem after taking the third or fourth cut from berseem. Rooted slips are planted in February (Central India) and in March (northern and north western parts) in lines by keeping a distance of 1 m between the rows and 30–40 cm between the plants. Berseem being an annual crop completes its life cycle in April and then the interrow spaces of hybrid Napier are prepared with a country plow and cowpea is sown in lines, 25-cm apart. In this way, in each set of two rows of hybrid Napier there will be two rows of cowpea. Cowpea is cut 60 days after sowing. Hybrid Napier continues to supply green fodder during the monsoon season. At the time of last cutting in October, the interrow spaces are again plowed up and the land is prepared for sowing berseem and Japan sarson to start the second cycle of the rotation. This cropping system ensures green fodder throughout the year. Hybrid Napier can be grown as sole crop as well as mixed crop with legumes such as cowpea, rice bean, etc. Hybrid Napier–legume mixtures are always desirable because of their complementary functions in providing nutritive, succulent, and palatable forage for dairy animals. Legumes as intercrop enhance the forage value and also add substantially the much nitrogen to the soil. The mixture also improves the physical condition of the soil, check soil erosion, resist the encroachment of weeds, and withstand the vagaries of weather better than pure stands.

10.14 NUTRIENT MANAGEMENT (MANURES AND FERTILIZERS)

Regular pattern of fertilizer application should always be followed for this high yielding grass. Thairu and Tessema (1985) reported that managing these high yielding forages without fertilizer is extremely impossible even in normal soil and rainfall conditions. Since this grass is a heavy yielder it requires high doses of nutrients. On the other hand, nutrient management has pronounced effect on yield and quality of the fodder grass. This grass responds very quickly to inorganic fertilization and livestock manure is an important resource for grass cultivation. These materials also help to improve soil's physical properties and increase the activity of beneficial soil microbes.

Application of NPK fertilizers should be done as per soil test values along with recommended farmyard manure or compost. In absence of soil test results 20–25 t farmyard manure should be applied during the time of land preparation and basal dose of 60 kg N, 50 kg P_2O_5, and 40 kg K_2O per hectare should be applied in bunds prior to planting. Walmsley et al., 1978 also suggested that highest yields could be expected from applying nitrogen after every cut. It has been observed that basal application of recommended NPK fertilizer has given higher establishment percentage and a higher yield of 359.7 t ha^{-1} $year^{-1}$. Subsequently 20 kg and 10 kg N should respectively be top dressed just after, and 20 days after the cut. Alternatively, the crop may be fertilized with 40 kg N just after the cut. Top dressing along with gentle raking of the soil produces more tillers.

10.15 WATER MANAGEMENT

The crop should be planted in well-moist soil condition. The first irrigation is done at the time of planting and immediately after planting, life irrigation should be given on the third day and thereafter once in 10 days. Frequencies of subsequent irrigation depend upon the rainfall and weather condition. The standard irrigation interval during summer is 10–12 days (depending upon the soil quality). The field should be provided with good drainage during the rainy season as the crop cannot stand water stagnation. Sewage or wastewater can also be used for irrigation.

10.16 WEED MANAGEMENT

Weeding should be done within 30 days of planting and second weeding is essential only if there is heavy weed growth. Hand weeding is done whenever necessary. Once in 2 years, intercropping with a suitable legume takes care of all the weeds. The legume, when intercropped with Napier, it suppress the weed growth, enriches the soil besides supplying nutritive fodder to the cattle.

10.17 DISEASE AND INSECT-PEST MANAGEMENT

No specific disease was noticed in hybrid Napier at present and except grass hopper, no major pest is recorded. Therefore, need-based disease and pest management should be taken.

10.18 HARVESTING AND YIELD

For efficient management and utilization of the crop, harvesting pattern is important. In a well-established crop, the first cut is ready in 60–65 days after planting and subsequent cutting are taken approximately at an interval of 25–30 days. Annually at least six to eight cuts are possible. If the crop gets more mature it becomes lignified and decreases its palatability and digestibility. Therefore, it is recommended to harvest the crop at right stage to feed animals and make silage if in excess. Generally it produces green forage yield of 150–300 t ha^{-1} $year^{-1}$. In order to encourage quicker regeneration from the basal buds, stubbles of 10–15 cm are to be left out at harvest.

10.19 SEED PRODUCTION

Seed setting is absent as it is an interspecific hybrid.

10.20 NUTRITIVE VALUE

Quality of forage grass is a vital parameter to ensure the fulfillment of all nutritional ingredients for animals. Herbage quality of hybrid Napier is very important as the leaves (foliages) are larger and greener, sheaths are softer, and margins are less serrated so the herbage is more palatable. It is juicer and succulent at all stages of growth. It is less fibrous and more acceptable. Premaratne and Premalal (2006) reported the average content of the nutritional components of the hybrid Napier variety CO-3 grown under average management condition, which is given below (Table 10.1).

TABLE 10.1 Nutritive Value.

Sr. No.	Nutritional ingredient	Value (dry matter basis) (%)
1.	Crude protein	15–16
2.	Crude fiber	34–37
3.	Acid detergent fiber	42–47
4.	Neutral detergent fiber	74–78
5.	Crude fat	6.2–6.9
6.	Ash	9.5–12.8
7.	Lignin	6–8
8.	Calcium	0.11
9.	Potassium	0.42
10.	Magnesium	0.36

These figures indicate the superior quality of the hybrid Napier var. CO-3 grass especially in crude protein, lignin, and fiber fraction than the other forage grasses.

10.21 UTILIZATIONS

It is one of the highest yielding perennial tropical fodder grasses and considered as cut-and-carry forage for stall feeder systems. The hybrid Napier once planted supplies fodder continuously and regularly for a period of 3 years. The cost of production is almost half that of single-cut crops. The production per unit area and time is approximately double than conventional fodders. Generally, if an animal is fed with adlib green forages of hybrid Napier that is grown under good management and cut at correct stage, it can be guaranteed that animal might take up most of the essential nutrients in sufficient quantities. Studies carried out on the nutritional aspect showed that this grass is more superior than most of the other farm grown fodder grass. To maximize the milk production of dairy animals and increase the profitability of dairy farming, hybrid Napier should be fed to the animals.

FIGURE 10.1 (See color insert.) Hybrid Napier.

KEYWORDS

- **hybrid Napier**
- **agronomic management**
- **green fodder**
- **nutrition**

REFERENCES

Bogdan, A. V. *Tropical Pasture and Fodder Plants*; Longman Group Limited: London, 1977, pp 233–234.

Premaratne, P.; Premlal, G. G. C. Hybrid Napier (*Pennisetum purpureum* × *Pennisetum americanum)* Var. CO-3: A Resourceful Fodder Grass for Dairy Development in Sri Lanka. *J. Agr. Sci.* **2006,** *2* (1), 22–33.

Thairu, D. M.; Tesema, S. Research on Animal Feed Resources; Medium Potential Areas of Kenya. In *Animal Feed Resources for Small Scale Livestock Producers*. Proceedings of the Second PANESA Workshop, Nairobi, Kenya, November 11–15. 1985, pp 127–137.

Walmsley, D.; Sargant, V. A. L.; Dookeran, M. Effect of Fertilizers on Growth and Composition of Elephant Grass (*Pennisetum purpureum*) in Tobago, West Indies. *Trop. Agric.* (Trinidad) **1978,** *25*, 329–334.

CHAPTER 11

GUINEA GRASS (GREEN PANIC GRASS)

SHYAMASHREE ROY[1*] and SANCHITA MONDAL GHOSH[2]

[1]*Regional Research Station, Old Alluvial Zone, Uttar Banga Krishi Viswavidyalaya, Majhian, Patiram, Dakshin Dinajpur, West Bengal, India*

[2]*Department of Agronomy, Bidhan Chandra Krishi Viswavidyalaya, Mohanpur, Nadia 741252, West Bengal, India*

[*]*Corresponding author. E-mail: shree.agr@gmail.com*

ABSTRACT

It is the most palatable fodder crop to animals because of its high leaf:stem ratio. It contains 10% crude protein at young leafy stage. Apart from its use as green fodder, it also can be used as hay or silage when the harvesting is delayed and the protein content starts decreasing. Guinea grass, also known as pasto guinea or mijo de guinea or green panic grass, is a perennial grass with high value as a palatable nutritious fodder crop. It is a large perennial bunch grass that is native to Africa, Palestine, and Yemen. It has been introduced in the tropic and subtropic around the world. The grass is very much preferred mainly for its taste, high yield, and good performance. Average green fodder yield is about 800–1000 quintals/ha per year. The production potential of guinea grass is 450–600 quintals/ha of green fodder per year in saline soils. In acid soils, the production potential of the grass is 800–900 quintals/ha. This grass contains hepatotoxin which causes photosensitization in animals. Animals with white skin or white patches usually suffer from this disease.

11.1 BOTANICAL CLASSIFICATION

Kingdom: Plantae
Order: Poales
Family: Poaceae
Genus: *Panicum*
Species: *maximum* (Jacq)
Binomial name: *Panicum maximum* (Jacq)

11.2 BOTANICAL NAME

Panicum maximum (Jacq) or *Megathhyrsus maximus* (Jacq).

11.3 COMMON NAME

Giiniigaas, gini ghaus, gini hullu, guinea grass, buffalo grass, pasto guinea, mijo de guinea, gini ghans, gewone buffel grass.

11.4 INTRODUCTION

Guinea grass, also known as pasto guinea or mijo de guinea or green panic grass, is a perennial grass with high value as a palatable and nutritious fodder crop. It is a large perennial bunch grass that is native to Africa, Palestine, and Yemen. It has been introduced in the tropic and subtropic around the world. The grass is very much preferred mainly for its taste, high yield, and good performance. It is the most palatable fodder crop to animals because of its high leaf:stem ratio. It contains 10% crude protein (CP) at young leafy stage. Apart from its use as green fodder, it also can be used as hay or silage when the harvesting is delayed and the protein content starts decreasing.

11.5 ORIGIN AND DISTRIBUTION

The crop was originated in tropical and subtropical Africa, but grows naturally in places, such as, Madagascar, Mauritius, and in some parts of Western Asia (Palestine and Yemen). Presently it is cultivated in different parts of the world, such as West Indies, parts of the United States, India, Sri Lanka, Malay, Australia, Philipines, New Guinea, etc. In India, the cultivation was initiated in 1793 and now it is cultivated throughout the country as a nutritious fodder (McCosker and Teitzel, 1975).

11.6 BOTANICAL DESCRIPTION

It is a perennial bunch-type grass with fibrous and deep root system. The plants may be as tall as 0.5–4.5 m., the stems may be of variable characteristics, strong, or delicate; rigidly upright, smooth, or covered with hair. The leaves may be 10–100-cm long and around 3.5-cm wide. *The genus Panicums* produces open flower heads (flowers in "loose panicles"), or much branched panicle with each spikelet containing a single floret. The small seeds are enclosed in a smooth hulls or glumes. Flowering and seed setting takes a very long time. The seeds are having shattering habit.

11.7 ECOLOGY

Warm and moist climate is best suited for guinea grass. The crop grows well to an altitude of 1800 m from mean sea level. But it is sensitive to frost. For growth of the grass, the suitable range for temperature is 15–38°C with a medium range of rainfall, that is, 700–900 mm. The grass prefers open grass lands and riverbanks of the tropical region and grows in such places naturally. Also, it loves the shaded places like coconut garden or underneath of any big trees. Therefore, it also can be used as a component of agroforestry system. Guinea grass has a deep root system which attributes to its tolerance to the drought. This grass cannot tolerate water logging and salinity. Few varieties can also thrive in low fertility and poor drainage condition (Middleton and McCosker, 1975). Two breeds, namely, "Vencedor" and "Centenario" are example of acidity tolerance.

11.8 SOIL

The grass can grow in any type of soil. A deep soil with moderate to high fertility is most preferred. Drainage facility is essential. Two to three ploughings are enough for land preparation.

11.9 SEED RATE AND SOWING

For sowing, seeds or rooted slips can be used as planting material. By using seed, a nursery is sown with 2.5–3.0-kg seeds and from 20 to 25-day-old nursery; seedlings are transplanted to the main field. But rate of germination is poor with seed. Therefore, vegetative propagation is preferred. For planting of 1 ha area, 25,000–66,000 rooted slips are sufficient. These are collected from aged clumps by uprooting them. For planting, field should be prepared in ridge and furrows. The slips are planted in furrows as three slips per hill.

11.10 SPACING

Spacing varies from place to place. The slips or seedlings may be placed at a spacing of 1 m × 0.5 m or 75 cm × 30–40 cm or 60 cm × 60 cm. In Kerala, when grown as intercrop, the spacing of 40 cm × 20 cm and a spacing of 60 cm × 30 cm for sole crop are followed, respectively. In North India, in case of legume intercropping, spacing is increased up to 1 m × 1 m or 2 m × 0.5 m in winter. 2–3 cm depth of sowing is followed (Maity and Mukherjee, 2008).

11.11 SOWING TIME

The crop is sown as rainfed with onset of southwest monsoon in the month of June. Otherwise as an irrigated crop, it can be sown at any time of the year. In North India, the best planting time is between mid-February and August.

11.12 CROP ASSOCIATION

Guinea grass, if grown as a sole crop, can leave adverse effects to the soil health in terms of soil nitrogen and soil organic matter. Therefore, association with legume crops is beneficial. Other than legumes, cereals or cole crops may also be included in the system. In North India, crops such as berseem, shaftal, senji, peas, oats, barley, Chinese cabbage, or Japanese rape are grown with guinea grass in winter and crops such as cowpea, guar, or rice bean are grown in summer. In monsoon, intercropping with lucerne may be followed.

11.13 VARIETIES

Bogdan (1955) classified 47 varieties of *Panicum maximum* into four broad groups:

a) Tall vigorous type: Robust plants with large leaves and rather thick stems. A fodder type of high productivity. For example, Coloniao, Hamil, and Coarse Guinea.
b) *Var trichoglume* type: Plants of medium vigor, with numerous fine stems. Leaves numerous, rather broad, and short. Basal and stem leaves are both numerous. Mainly a grazing type.
c) Medium sized type: Narrow, predominantly basal leaves. For example, Makueni.
d) Annual type.

The species *Panicum maximum* is highly variable taxonomically and phytogenetically. The commercially available *Panicum maximum* can be broadly grouped in two distinct types: the guineas and the panics. Guineas are produced mostly on high-rainfall tropical lowlands, whereas panics are grown in different parts of the subtropics or elevated moist tropics. The panic group includes Petrie, Sabi, and Gatton.

Depending upon the agronomic characters, the species can be grouped into two types: large or medium type and small or low growing type. Small type guinea grass, green panic or slender guinea (var. trichoglume) named cultivar "Petrie" from Australia is important.

Commercially important varieties for large or medium type from different countries:

a. Queensland common (East Africa)
b. Riversdale (Riversdale, Australia)
c. Makueni (Kenya)
d. Gatton panic (Africa)
e. Hamil grass (North Queensland)
f. Coloniao guinea (Barbados)
g. Sabi (Rhodesia)

h. Embu (Kenya)
i. Punjab guinea grass (Australia)
j. Borinquen (Puerto Rico)
k. Broad-leaf (Puerto Rico)
l. Semper varde (Brazil)
m. Sigor (Kenya)
n. Nichisi (Kenya)
o. Kingranch (The United States)

11.14 NUTRIENT MANAGEMENT

This forage grass performs better when grown in organic matter enriched soil. 20–25 t FYM/ha should be mixed well at the time of land preparation. Dry matter yield of guinea increased linearly with per ton increase of the farm yard manure (FYM) applied. Since, animal production is directly related to forage quality, information is therefore required on the role of FYM application on forage quality. Ahmed et al. (2012) reported that CP concentration of guinea increased linearly with FYM incorporation. Inorganic fertilizers (N, P_2O_5, and K_2O) should be applied based on the soil test values. In absence of soil test report, the crop may be fertilized with the recommended fertilizer dose, that is, 60 kg/ha N, 50 kg/ha P_2O_5, and 40 kg/ha K_2O as basal dose in bands at the time of sowing. Subsequently, 40 kg/ha nitrogen should be top dressed after each cut. The critical dietary CP level in tropical forages is 7%, below which voluntary intake is down. The nitrogen @ 60–100 kg/ha every 40−45 days were needed to produce CP concentrations in leaves of guinea grass above 7%. At lower rates of 20–40 kg N/ha, rapid growth of guinea plants was observed due to a dilution effect, so that CP concentrations in leaves were similar to the plants receiving no nitrogen.

The use of inorganic P and K fertilizers resulted in higher yield and better quality of forage at the first cut but they were not significantly different from the control (without P and K) at the fourth cut (Ahmed et al., 2012). In calcareous soils, this crop often suffers from sulfur deficiency and application of 50 kg S/ha is recommended for higher biomass production and quality improvement. This grass benefits well from biofertilizer (*Azotobactor* + *Azoapirillum*) as these biofertilizers help to make the phosphate available in soil (Mohammadi, 2012; Sharma et al., 2013) and guinea grass responds very well to phosphorus fertilizers.

11.15 WATER MANAGEMENT

Two irrigations are necessary within 7–10 days of planting for quick establishment. The subsequent irrigations should be given depending upon the soil type and weather condition. Usually irrigation once in 7–10 days is needed. Irrigation with cowshed washing or sewage water within 3–4 days after each cutting provides better production.

11.16 WEED MANAGEMENT

The delicate seedlings or newly emerged shoots from slips or cuttings should be protected from weeds in the first 2 months. Two hand weeding/hoeing should be given during this period, which will ensure good aeration and crop growth as well as check weeds. Later, intercultivation may be necessary after three or four cuttings.

11.17 DISEASE AND INSECT-PEST MANAGEMENT

Generally guinea grass is not much affected by any insect-pest and disease but leaf spot (*Bipolaris hawaiiensis*) is often found on leaves during the wet season. There is no evidence that this disease affects production.

11.18 HARVESTING

The grass is harvested directly through grazing or by machine for green chop, silage, or hay. For optimal forage quantity and quality, to know when and how to harvest, an understanding of grass growth and regrowth mechanism is very essential (Waller et al., 1985). The crop is ready for harvest when it reaches 1.5-m height. Cutting at 15–20 cm above the ground level is advised. The first cut is usually ready in 60–65 days after planting and subsequent cuts are taken at 25–30 days intervals. But, cuttings taken at 30 and 35 days interval seem to provide maximum yield with an acceptable leaf to stem ratio and CP content. About six to eight cuts can be made in a year. In order to encourage quicker regeneration from the basal buds, stubbles of 10–15 cm are to be left at harvest. Nandi et al. (2015) also reported that harvest at either 10 or 15 cm produced higher growth and establishment of the grass.

11.19 YIELD

Average 800–1000 quintals/ha of green fodder is obtained per year. But, the production potential of guinea grass is 450–600 quintals/ha of green fodder per year in saline soils. In acid soils, the production potential of the grass (Hamil, PGG-9, and PGG-14) is 800–900 quintals/ha.

11.20 USE

It is the most palatable fodder crop to animals because of its high leaf:stem ratio. It contains 10% CP at young leafy stage. Apart from its use as green fodder, it also can be used as hay or silage when the harvesting is delayed and the protein content starts decreasing.

11.21 NUTRITIVE VALUE

The CP content of the young grass, mature grass, and the straw offered were 12.2%, 5.4%, and 7.7%, respectively. The crude protein (CP) content of the grass ranges from 4% to 14%. But some varieties content even 20% CPs at around 28 days age. The per cent of CP decreases with age of the plant. The organic matter digestibility of the young grass (69%) was significantly higher than the mature grass (62.5%) and straw (55.8%) diets.

11.22 TOXICITY

The species *Panicum* contains hepatotoxin which causes photosensitization in animals. The affected animals may be fed on chlorophyll-free diet or kept in darkness for a few days till recovery. Animals with white skin or white patches usually suffer from this disease (Maity and Mukherjee, 2008). A photosensitization disease, known as Dikoor in sheep, reported from South Africa. The cultivar "Petrie" is reported to cause hyperparathyroidism in horses, and nephrosis or hypocalcemia is reported due to oxalate accumulation.

FIGURE 11.1 (See color insert.) Guinea grass.

KEYWORDS

- **guinea grass**
- **agronomic practices**
- **yield**
- **quality**
- **toxicity**

REFERENCES

Ahmed, S. A.; Halim, R. A.; Ramlan, M. F. Evaluation of the Use of Farmyard Manure on a Guinea Grass (*Panicum maximum*)–Stylo (*Stylosanthes guianensis*) Mixed Pasture. *Pertanika. J. Trop. Agric. Sci.* **2012,** *35* (1), 55–65.

Bogdan, A. V. Herbage Plants at the Grassland Research Station, Kitale, Kenya. *E. Afr. Agr. Forestry J.* **1955,** *30*, 330–338.
Maiti, S.; Mukherjee, A. K. Forage Crops Production and Conservation; Kalyani Publisher: New Delhi, 2008.
McCosker, T. H.; Teitzel, J. K. A Review of Guinea Grass (*Panicum maximum*) for the Wet Tropics of Australia. *Trop. Grassl.* **1975,** *9* (3), 177–190.
Middleton, C. H.; McCosker, T. H. Makueni—A New Guinea Grass for North Queensland. *Qld. J. Agr. Anim. Sci.* **1975,** *101*, 351–355.
Mohammadi, K. Phosphorus Solubilizing Bacteria: Occurrence, Mechanisms and Their Role in Crop Production. *Resour. Environ.* **2012,** *2* (1), 80–85.
Nandi, C. C.; Onyeonagu, C. C.; Eze, S. C. Growth Response of Guinea Grass (*Panicum maximum*) to Cutting Height and Poultry Manure. *Am. J. Exp. Agr.* **2015,** *7* (6), 373–381.
Sharma, S. B; Sayed, R. Z.; Trivedi, M. H.; Gobi, T. A. Phosphate Solubilizing Microbes: Sustainable Approach for Managing Phosphorus Deficiency in Agricultural Soils. *SpringerPlus* **2013,** *2*, 587.
Waller, S; Moser, L.; Reece, P. *Understanding Grass Growth: The Key to Profitable Livestock Production*; Trabon printing Co., Inc.: Kansas City, Missouri, 1985.

PART II
Leguminous Forages

CHAPTER 12

COWPEA (BLACK-EYED PEA)

ABHIJIT SAHA[1*], SONALI BISWAS[2], DULAL CHANDRA ROY[3], and UTPAL GIRI[1]

[1]*Department of Agronomy, College of Agriculture, Lembucherra 799210, Tripura, India*

[2]*Department of Agronomy, BCKV, Mohanpur 741252, Nadia, West Bengal, India*

[3]*Department of ILFC, WBUAFS, Mohanpur 741252, Nadia, West Bengal, India*

[*]*Corresponding author. E-mail: abhijitsaha80@gmail.com*

ABSTRACT

Cowpea is most potential legume for fodder production, crucial for crop–livestock integration, enrichment of soil, conservation of soil moisture, and preventing soil erosion. Fodder cowpea is a fodder legume inherently more tolerant to drought than other fodder legumes. Cowpea is a quick-growing crop, which produces tremendous quantity of bulk mass in short span of time; and therefore, is esteemed as valuable catch crop or fodder, green manure, grains, or vegetable. The green fodder yield of 350–400 quintals/ha are obtained from multicut types, and about 250–300 quintals/ha from single cut types 50 quintals/ha of dry matter yield can be achievable. It has about 16% crude protein and 20% crude fiber. Cowpea contains some antinutritional factors which create some physiological disorders on consumption. Some of them are protease inhibitors, antivitamins, phytase, saponins, amylase inhibitors, tannins, aflatoxins, etc.

12.1 BOTANICAL CLASSIFICATION

Kingdom: Plantae
Orders: Fabales
Family: Fabaceae
Subfamily: Faboideae
Tribe: Vicieae
Genus: *Vigna*
Species: *unguiculata* (L.) Walp
Binomial name: *Vigna unguiculata* (L.) Walp

12.2 BOTANICAL NAME

Vigna unguiculata (L.) Walp

Synonyms: *Vigna sinensis* (L.) Savi ex Hassk; *Vigna catjang* (Burm. f.) Walp

12.3 COMMON NAMES

English—cowpea, bachapin bean, cherry bean, black-eye bean or pea, catjang, china pea, cowgram, and southern pea; Hindi—barbati, lobia, alsande, and chawli

12.4 INTRODUCTION

Increased urbanization, income, and changed lifestyle in developing countries may considerably increase the demands of livestock products such as milk and meat. Poor nutrition remains the most widespread technical constraint to good animal performance in India. By and large, livestock production in India is done by small holders, characterized by lower production of milk, growth, and extended calving period. This becomes more critical during the dry season when feed availability is not only inadequate but also the quality extremely poor. Integration of forage legumes into the feeding strategies is the most promising solution of this problem (Nnadi and Haque, 1988). Although feeding of forage legumes has been found easily adoptable, the practice is not attractive to the farmers. Farmers do not pay particular attention to the planting of pure legume stands rather greater emphasis is on the cultivation of food crops. In an integrated crop–livestock farming systems of India, planting of dual-purpose legumes such as cowpea is gaining immense popularity. Its fodder is a good dry season feed supplement to livestock. Cowpea has a significant contribution of total green fodder production in India and can maintain milk yield of 15 gallons/cow/day by giving dry matter yields of more than 18 t/acre and protein content up to 26.0%.

Cowpea is, therefore, the most potential legume for fodder production, crucial for crop–livestock integration, enrichment of soil, conservation of soil moisture, and preventing soil erosion. Fodder cowpea (*Vigna unguiculata* L. Walp) is a fodder legume inherently more tolerant to drought than other fodder legumes, and considered as a crop capable of improving sustainability of livestock production through its contribution in improving seasonal fodder productivity and nutritive value. The first written reference using cowpea appeared in 1798 in the United States. The name was most likely acquired due to their use as a fodder crop for cows. The common name of black-eyed pea, used for the *unguiculata* cultivar group, describes the presence of a distinctive black spot at the hilum of the seed. Black-eyed peas were first introduced to the southern states in the United States and some early varieties had peas squashed closely together in their pods, leading to the other common names of southern pea and crowder pea. Sesquipedalis in Latin means "foot and a half long," and this subspecies which arrived in the United States via Asia is characterized by unusually long pods, leading to the common names of yard-long bean, asparagus bean, and Chinese long bean.

12.5 TAXONOMY

The cowpea belongs to genus *Vigna*, comprising many pulse crops of the old world. This species contains two wild (*dekindtiana* and *mensensis*) and three cultivated (*unguiculata*, *cylindrical*, and *sesquipedalis*) forms.

12.6 ORIGIN AND DISTRIBUTION

Cowpea (*V. unguiculata* L. Walp) is one of the most ancient human food sources and has probably been used as a crop plant since Neolithic times. As postulated by Ng (1995) that during the evolution process of *V. unguiculata*, there was a change in its growth life cycle, from perennial to annual. However, the wide geographical distribution of the wild species throughout sub-Saharan Africa, especially West Africa, indicates that they could have been domesticated in any part of Africa. The distribution of wild forms covers much of tropical Africa, whereas the greater part of variability within the wild species is confined to South Africa. The greatest genetic diversity of cultivated cowpea is found in West Africa, in the savanna region of Burkina Faso, Ghana, Togo, Benin, Niger, Nigeria, and Cameroon. Some literature indicates that cowpea was introduced from Africa to the Indian subcontinent approximately 2000–3500 years ago.

It is estimated that the annual world cowpea crop-growing area is 12.5 million hectare, and the total grain production is 3 million tons. West and Central Africa is the leading cowpea producing region in the world. This region produces 64% of the total estimated 3 million tons of cowpea seed produced annually, having area of 8 million hectares. Nigeria is the world's leading cowpea producing country, followed by Brazil. Other countries in Africa, for example, West Africa, are Nigeria, Senegal, Ghana, Mali, and Burkina Faso. Ghana, Niger, and Cameroon are significant producers. The major production areas elsewhere in the world are Asia (India and Myanmar) (about 1.3 million hectares) and the Central and South America (United States, Brazil, and West Indies) (2.4 million hectares). In India, fodder cowpea is cultivated in 3 lakh hectares area with a green fodder productivity of 25–45 t/ha.

12.7 PLANT STRUCTURE

Cowpea is a vigorously growing herbaceous annual herb, bushy, trailing, or climbing. Most cowpea varieties are, in general, sensitive to photoperiodic induction. They are generally quantitative short-day plants with tendency to flower as the day lengths become shorter. The day length influencing flowering may vary due to differential response of varieties, but it lies close to 13.5 h. Cowpea has a strong taproot and many spreading lateral roots in surface soil.

12.7.1 LEAVES

The first pair of leaves is basic and opposite while the rest are arranged in alternate patterns and are trifoliate. The leaves are usually dark green in color. Leaves exhibit considerable variation in size (6–16 cm × 4–11 cm) and shape (linear-lanceolate to ovate). The leaf petiole is 5–25-cm long.

12.7.2 STEM

Striate, smooth, or slightly hairy with some purple shades.

12.7.3 INFLORESCENCE

Flowers are arranged in racemose or intermediate inflorescences at the distal ends of 5–60-cm long peduncles. Flowers are borne in alternate pairs, with usually only two to a few flowers per inflorescence. Flowers are conspicuous, self-pollinating, borne on short pedicels, and the corollas may be white, dirty yellow, pink, pale blue, or purple in color. They are mostly self-pollinated, but outcrossing may occur only 2%.

12.7.4 FRUIT AND SEED

Pods are pendulous, smooth, 10–23-cm long, and cylindrical with a thick decurved beak often containing 10–15 seeds. Usually yellow when ripe, but may also be brown or purple in color. Seeds vary considerably in size, shape, and color. Usually the number of seeds per pod may vary from 8 to 20. The seeds are relatively large (2–12-mm long) and weigh 5–30 g/100 seeds. The testa may be smooth or wrinkled; white, green, buff, red, brown, black, speckled, blotched, eyed (hilum white, surrounded by a dark ring), or mottled in color. Usually yellow when ripe, but may also be brown or purple in color.

12.8 CLIMATE AND SOIL

Cowpea is highly suited to moderately humid areas of tropics and subtropics. It generally grows in areas between 30°N and S, up to 1500 m elevation. The plants are unable to withstand frosts, excessive heat, and prolonged waterlogging. It is grown as a summer or *kharif* crop in high monsoon areas. The base temperature for germination is 8.5°C and for leaf growth 20°C. Cowpea is a heat-loving and drought-tolerant crop. The optimum temperature for growth and development is around 26–30°C and the minimum is 15°C. Therefore, cowpea can be grown as winter crop in the eastern and southern regions of India. For forage cowpea, well-distributed rainfall of 750–1100 mm is preferable. It will tolerate lower rainfall, but in high rainfall areas disease and insect attacks increase. Intermittent showers followed by bright sunshine are highly conducive for cowpea growth.

Cowpea grows well in a wide range of soil, from well-drained heavy clay soil to varying proportions of clay and sand. It is more tolerant to infertile and acid soils than many other crops. The crop thrives best in slightly acid to slightly alkaline pH (5.5–8.0), sandy loam soils. Cowpea shows little tolerance to salinity but tolerates soils high in aluminum. Saline, alkali, or waterlogged soils must be avoided.

12.9 LAND PREPARATION

The land should not be waterlogged, but well drained. During land preparation, the existing perennial weeds, trees, and shrubs in the site are cut down manually, or slashed with a tractor. This should be followed by two to three plowing with a desi plow and two to three harrowings using a disc plow and harrow. Some 4–6 days between each operation should be allowed to enhance good soil tilth and bring the land in optimum condition for good seed germination. The land may be ridged in high rainfall areas or left as flat seedbeds after harrowing, keeping a mild slope of the field to drain out excess water. In irrigated crop, shallow furrows at 3 m apart for leading irrigation water may also be provided.

12.10 SOWING

Cowpea is directly grown from seed. Cowpea does not suffer from seed dormancy, but it possesses epigeal type of germination. Seed rate can be estimated using the following formula:

$$\text{Seed rate} = \frac{\text{Plant population required}}{\text{Number of seeds/kg} \times \text{germination\%} \times \text{establishment\%}}$$

Optimum soil temperature for rapid germination of cowpea seed is above 18.3°C.

12.10.1 SOWING TIME

Fodder cowpea is not season bound and can be grown in any month of the year, provided irrigation facilities are available, but very hot summer and severe winter must be avoided. In northern India, cowpea is sown from March to August. In the eastern India, rainfed cowpea is sown with the onset of monsoon in June–July. In Kerala, fodder cowpea is raised as a rainfed crop during May–June and as a summer crop in rice fallows after harvesting of second crop of rice on December–January. In West Bengal condition, winter cowpea is not grown because severe winter temperatures cause stunted growth of the plants and so, here it is normally sown in June–July but it can be sown in March–April also with adequate irrigation facility. Generally for the fodder purpose, early sowing is much preferable.

12.10.2 SOWING METHOD

Cowpea can be sown by broadcast, drilled, or dibbled in lines depending on the purpose or season. Line sowing has been better over broadcasting method of sowing. However, for fodder and green-manuring broadcasting method is generally considered better. Ridge–furrow system has been superior over flat system of sowing in Maharashtra and Kerala. In Kerala conditions, dibbling two seeds per hole were practised for grain and dual-purpose cowpea. In high rainfall zone, 30-cm wide, 15-cm deep, and 2-cm apart channels are made for draining of excess rain water. Seeds germinate fast when sown in proper moisture at a depth of 4–5 cm. When grown as mixed crop with cereal fodders, the two crops may be sown in one or two lines alternating each other with reduced seed rates or in crop-wise direction at their full normal rates of sowing.

12.10.3 SPACING

Inter row spacing is utmost important for cowpea due to busy, trailing, and erect type varieties being used for different purposes and in different seasons. For bushy and dwarf varieties, close spacing of 30 cm may be adopted, whereas for semi spreading types 40–45-cm spacing is used. In Bangalore and Pattambi, closer spacing of 30 cm was observed superior to wider spacing, whereas opposite may be the case in rainfed arid regions. Closer spacing may be preferred in late planting in *kharif*. In spring and summer season due to lesser vegetative growth, closer spacing of 25–30 cm may be kept. Generally for fodder cowpea production in West Bengal, row to row spacing of 30–40 cm along with 10–15-cm plant to plant spacing is maintained.

12.10.4 SEED RATE

Seed rate is quite variable factor and depends on sowing time, type of soil, seed size, purpose, and soil moisture availability. Recommended seed rate is 50 kg/ha for a broadcasted crop and 35–40 kg/ha for a drilled or dibbled crop. For grain purpose, seed rate should be 15–20 kg/ha, whereas seed rate for intercropping with cereal forage may vary from 15 to 20 kg/ha.

12.11 CROP MIXTURES

Cowpea is tolerant of moderate shade, so it is successfully grown with other crops such as maize, jowar, bajra, teosinte, Sudan grass, hybrid Napier, guinea grass, Deenanath grass, etc. Maize and sorghum are mixed for preparing silage, while those with Sudan grass and sorghum for making hay. The crop mixtures usually produce high yield and huge returns. It can also be grown with pasture grasses such as Anjan, Marvel, Schima, and Sewan.

12.12 NUTRIENT MANAGEMENT

Cowpea can be grown without any manure or fertilizer in nutrient-rich soils. Nodulation occurs freely with native *Rhizobia* and fixes considerable amount of nitrogen. Cowpea is nonspecific (promiscuous), but inoculation with selected cowpea strain enhances the nodulation, increases the dry matter production, and increases uptake of K and Ca. High soil temperature also affect nodulation, but deep planting helps to overcome this problem. In low-fertile soils, cowpea responds favorably to calcium where pH is low; but this may be due to the effect of released molybdenum, which is essential for nitrogen-fixing bacteria.

Generally on nutrient-deficient degraded soil, 5–15 t/ha farm yard manure (FYM) may be applied. However, long-term fertility experiments revealed that optimum dose of chemical fertilizer along with FYM may be much useful than the fertilizer alone. Das et al. (2011) reported that growth parameters such as plant height, number of leaves, and branches per plant along with all yield-contributing characters and yield were significantly increased to a greater extent by the treatment of 75% recommended dose of fertilizer (RDF) + vermicompost + *Rhizobium* + phosphorus solubilizing bacteria (PSB) as compared to RDF alone. It indicates a saving of 25% chemical fertilizer. Vermicompost may increase yield of cowpea by 23% over FYM @ 20 t/ha, followed by vermicompost @ 10 t/ha. Vesicular arbuscular micorrhizae fungus offers promise in cowpea as it enables the inoculated plants to uptake more effectively the available phosphorus from the soil, also there is enhanced absorption of water, nitrogen (N), potash (K), and micronutrients. Abdel-Aziz and Salem (2013) reported that seed inoculations with individual *Azospirillum* sp. and *Trichoderma* sp. obtained the superiority treatment for vegetative growth characters and nitrogen (%) in leaves and protein content (%) in seeds. PSB biofertilizer helps in increased phosphorus (P) availability to cowpea.

The application of nitrogen in cowpea basically depends on the status of organic carbon of the soil. In case organic matter in the soil is below 0.5%, full dose of nitrogen should be applied and if it is between 0.5% and 2.0%, the dose may be accordingly modified. The usual fertilization of cowpea should consist of 20 kg N, 40 kg P_2O_5, and 20 kg K_2O per hectare, and they are all to be applied as basal. Fertilizer at sowing may be drilled 5–7 below the seed zone. However, for irrigated crop, fertilizers shall be applied in two splits at an increased level. For this, the basal dose of NPK should be given @ 40:30:30 kg/ha. Topdressing of N and K each at 10 kg/ha shall also be provided after each cut. Proper P application helps in root development and microbial activity for increased absorption of mineral salts. Mandal et al. (2009) pointed out that highest cowpea yields were obtained when N, P, and K were applied at rates 25% higher than the soil test-based optimum rate, keeping S, Zn, and B at their optimum levels. Choudhary and Suresh Kumar (2013) reported that higher level of K (40 kg/ha) and P (60 kg/ha) helped efficient production as well as utilization of energy and economic benefit. Under P deficiency, synthesis of protein is adversely affected and arginine accumulation results in restricted plant growth. Cowpea performs well to spraying of 3% solution of superphosphate 5–6 weeks after sowing. Potassium has been lauded in cowpea for promoting growth and mitigating drought effects during suboptimal soil water stress.

Calcium is equally important for cowpea, particularly, for slightly alkaline and acid soil, hence gypsum on the former and calcium carbonate on the latter soil @ 200–300 kg/ha may be applied as basal dose at sowing. Sulfur is more important in cowpea than nitrogen, governing the yield and quality of fodder of cowpea. Since boron, molybdenum, copper, zinc, and manganese are important for nodulation, these minerals should be applied in small amount whenever there is any deficiency of those in soil. From the data shown in the following table, it can be concluded that 20 kg N + 40 kg P_2O_5/ha along with *Rhizobium* seed inoculation gave significantly higher growth, maximum nodulation, and seed yield of cowpea cv. pusa phalguni in Navsari Agriculture University (NAU), Gujarat during summer season (Patel and Jadav, 2010).

12.13 WATER MANAGEMENT

A presowing irrigation is important for the proper germination of the crop. Follow-up irrigations may be given at every 20–25 days in summer and at monthly intervals during the postmonsoon period. Proper drainage must be provided to avoid waterlogging. Consumptive use of water with increased irrigation levels was highest (44.8 cm) at 7.5 cm cumulative pan evaporation (CPE) and lowest (13.4 m) without irrigation at Pantnagar. Water use efficiency was highest (90 kg green fodder/ha/mm) in the variety UPC 5286. The results of the experiment done by Anita et al. (2015) revealed that plant height and number of branches were maximum at irrigation at irrigation water (IW)/CPE ratio of 0.8 and among the varieties, COFC–8 has presented a more important morphological growth than other varieties, which in turn leads to relative tolerance of this cultivar under moisture stress situations. The fodder cowpea fertilized with 40 kg P_2O_5/ha has improved the stress tolerance of cowpea when not irrigated as plants fertilized with high levels of P not only survive but also are able to hold great potential leaf extract water from deep soil, allowing satisfactory return. Khalil et al. (2015) inferred on the basis of experimental results that under moderate water stress 75% of the commonly practiced irrigation, application of 60 kg K_2O fed^{-1}, could sustain cowpea plant to grow satisfactorily.

Jha et al. (2016) showed that drip irrigation in fodder cowpea used 73% less water while yielding 7% more dry matter. Water use efficiency was 4.42 kg/m^3 for drip irrigation compared to 1.45 kg/m^3 for furrow irrigation in Kapilvastu, Nepal.

12.14 WEED MANAGEMENT

Muhammad et al. (2003) reported that the presence of weeds in cowpea reduced yield by 82% and a significant increase in yield of pods was noted by controlling weeds up to 45 days of sowing. In other words, cowpea should be kept free of weeds after establishment. So, for this one weeding at 20–25 DAS may be required in early growth stages to combat the weed problems. However, if weeds still develop due to repeated showers, one more weeding 20–25 days later may be done, very high seed rates are often effective to smother the weeds in forage crops.

Manual weeding is becoming difficult day by day due to lack of labors and time. So, herbicides can be used effectively to control early flush of weeds. A preemergence application of pendimenthalin at 0.75 kg a. i. ha^{-1} + weeding at 5 weeks after planting significantly gave higher grain yield of 511 kg ha^{-1} and net return of $4705 ha^{-1} compared to other treatments. Parasuraman (2000) found that application of pendimenthalin (1.5–2.0 L ha^{-1}) or fluchloralin (1.0–1.5 L ha^{-1}) at 3 days after planting (DAP) + hand weeding twice at 30 DAP resulted in a significant reduction in weed population, weed dry matter, and an increase in cowpea yield under rainfed conditions. Madukwe et. al. (2012) reported that chemical weeding at 2–3 leaf stage of weed + hand weeding at 50 DAP proved effective weed control method in cowpea fields. According to Muhammad et al. (2003), the best postemergence herbicide for the control of weeds in cowpea is phenoxaprop-p-ethyl at the rate of 80 g ha^{-1} + glyphosate at 1800 g ha^{-1}, and was observed to be more effective against grasses.

12.15 HARVESTING

Fodder cowpea can be utilized in many ways. It can be grazed, cut and fed as green fodder, or be made silage and hay. Grazing or cutting should not be allowed before flowering as regeneration rate is very low. For quick regeneration, four to six buds per plant must be left during cutting. Cowpea can be harvested at 50% flowering. In a two cut management, the first cut is taken at 50–55 days at 15 cm above ground level, while the second cut is taken after 40–45 days of the first cut. Generally, the growth of cowpea ceases beyond the leaf area index of four or six, and so it is better to harvest the crop at that stage. Cowpea makes good hay, but care must be taken to preserve the leaf. Harvesting for hay is done when most of the pods are full-grown and a considerable number of them are mature.

12.16 YIELD AND FODDER QUALITY

The green fodder yields of 350–400 quintals/ha is obtained from multicut types, and about 250–300 quintals/ha from single cut types; 50 quintals/ha of dry matter yield can be achievable. It has about 16% crude protein and 20% crude fiber. The nutritive value of fresh biomass of cowpea (DM basis) is almost 125%, digestible crude protein (62.0%) total digestible nutrition 2.7 M cal/kg digestible energy, and 2.2 M cal/kg metabolizable energy. The corresponding values for its hay and seed are 102%, 53.0%, 2.4%, and 1.9% and 21.1%, 84.0%, 3.7%, and 3.3%, respectively. It is usually ensiled mixed with sorghum or maize in the ratio of one part of cowpea and two parts of maize or sorghum to provide sugar for fermentation, to add bulk, and to prevent protein losses. Cowpea makes good hay, but care must be taken to preserve the leaf.

12.17 SEED PRODUCTION

Cowpea is a self-pollinated short-day plant. Because of uneven ripening of seeds, harvesting is difficult. In the case of prostrate or trailing types, the crop will be cut or sown when two-thirds of the pods are dry and rattle when shaken. The vines should be thoroughly dried before threshing. As the problem of uneven ripening, hand picking of pods many times in the season will give the highest yields. The percentage of hard seeds is low and seeds remain viable for 3 years under storage. Seed certification standards for cowpea are: minimum germination 75%; minimum pure crop seeds 98%; maximum weed seeds 0.1%; maximum other crop seeds 0.05%; and maximum inert matter 2.0%.

12.18 DISEASE MANAGEMENT

The major diseases account for more than 20 viral, 30 fungal, 10 bacterial and nematode origins. All diseases are not equally devastating but their intensity greatly changes. Many of the diseases are seed/soil born in nature, but most destructive ones are viral ones and may be transmitted from one generation to next through seed, and thus disseminated to most cowpea-producing regions of the country.

12.18.1 ANTHRACNOSE

The symptoms of this disease are leaf spot, stem blackening, pod discoloration, seed rot, and seedling blight. To control anthracnose, seed treatment with Mancozeb, benomyl, and Thiram @ 0.2% a.i. spray at fortnight interval is recommended.

12.18.2 CHARCOAL ROT/SEEDLING

The symptoms are stunted growth and rot in hypocotyls and roots. Seedling may dry up. In mature cowpea, dry or wet greyish black sunken lesions on the lower stem. To control, 15% neem cake was most effective in reducing disease infection. Combined treatment of bavistin at the rate of 1 g/kg seed and foliar spray of blitox-50 (0.03%) performed best.

12.18.3 FUSARIUM WILT

The symptoms are stunted growth, chlorosis, droping, withering of leaves, and venial necrosis. To control fusarium wilt, seed treatment with Thiram (0.3%), Captan (0.2%) singly or in combination and application of biocontrol agents such as *Trichoderma harzianum* and *Trichoderma viride* is recommended.

12.18.4 CERCOSPORA LEAF SPOT

Rough circular cherty red to dark red spots on the upper leaf surface; leaves may turn yellow leading to defoliation, and on the lower leaf surface lesions are red. To control, seed treatment with Thiram or captan @ 2.5 g/kg and foliar application of Dithane M-45 (0.2% a.i.) or Dithane Z-78 (0 2% a.i.) is recommended.

12.18.5 POWDERY MILDEW

Small white patches on the upper surface of the leaves, petioles, branches, stem, and pods. Affected plants remain short stature; leaves turn yellowish, curl, wither, and die. Spray crop with wettable sulfur (0.2%) or benomyl (0.2%) and application of 0.05% Kelthane as spray immediately on appearance of disease and twice at 15 days interval is recommended.

12.18.6 BACTERIAL BLIGHT

Water-soaked dots having size of 0.5–2.0 cm which appear on the lower surface of the leaf. Complete defoliation may occur, and the infected stems become cankerous near the ground level and cracking occurs, stunted and bushy in appearance. Hot water treatment at 52°C for 15 min. Treatments with mercuric chloride, antibiotics, solar heat treatments, and hot water treatment for surface sterilization of infected seeds.

12.18.7 COWPEA YELLOW MOSAIC

Infected plants exhibit yellow mosaic foliage symptoms. Leaf yellowing is accompanied by leaflet deformation, dwarfing, and plant stunting. Weeds acting as alternative host should be removed and destroyed. Rouging of infected plants just after appearance of symptoms. Spraying the crop with

0.2% endosulfan may check the insect vectors. However, application of insecticides may not achieve complete control of the disease.

12.19 INSECT-PEST MANAGEMENT

Almost 85 insect species cause infestation in cowpea. It is, therefore, desired to identify the symptoms, extent of losses, and to take preventive and control measures against major insect-pests so that yield levels by preventing or lessening their impacts could be enhanced.

12.19.1 WHITEFLIES

Feeds on the plants and also transmits yellow mosaic virus), leaf turns yellow, destroy chlorophyll of leaves, plants growth is stopped, and the loss may range from 12% to 65%. Monocrotophos @ 0.25 kg a.i./ha applied three times at 15 days interval starting from 15 days of sowing. Application of phorate @ 1–1.25 kg ha at sowing keeps the whitefly population checked up to 4 weeks of plant growth. Dimethoate @ 0.04% is also effective.

12.19.2 JASSIDS

A sucking pest, sucks leaf sap, leaflets become cup shape and yellow at the edges. In heavy infestation, leaves become brown and dry up. Application of seed coating of aldicarb @ 1.0 kg a.i./ha followed by monocrotophos spray in 3–5-week-old cowpea crop.

12.19.3 APHIDS

Results twisted young seedling, and on heavy infestation, seedlings may show mortality. Due to sucking of sap, plant growth is stopped. Yield loss up to 44–87%. Dimethoate @ 0.04% has been observed most effective. Effective control can be done by spraying endosulphan (0.07%), monocrotophos (0.04%). or lindane (0.01%). Some natural enemies of aphids are lady birds (*Coccinella septempunctata*) and syrphid larvae (*Ischzeden scutellons*).

12.19.4 WEEVILS

Chew the leaflets generally at the margin and make central hole grubs feed in soil and crop roots causing destruction of roots. The adults feed on leaves, buds, flowers, and young pods. Spraying of Malathion dust @ 20–25 kg/ha.

12.19.5 TERMITES

The damage increases under drought situations or the interval between two irrigations/rains. Mixing of soil insecticides such as chlorophyriphos @ 25 kg/ha of dust formulation before sowing of the crop.

12.20 INTEGRATED PEST MANAGEMENT

Integration of early sowing of the crop, use of appropriate agronomic practices, insecticides, plant extracts, and cultivating resistant varieties may lead to effective control of major pests of fodder cowpea.

- Planting of crop in first or second week of July is considered most profitable with increased yield and reduced pests populations.
- Use of aqueous neem, neem kernel extracts, and neem seed powder may provide complete pest protection to fodder cowpea.
- Initial application of neem seeds pellets @ 200 kg/ha at the time of sowing placed in furrows may check the crop against termite and white grub.
- Neem seed extract acts as repellent and growth inhibitors against red hairy caterpillar.

12.21 VARIETIES

The important forage cowpea varieties are Kohinoor (S 450), HFC-42-1, GFC-1, GFC-2, GFC-3, GFC-4, UPC-5286, CO-5, UPC-5287, Sweta, Charodi, UPC-287, Gujarat copea-3, UPC-4200, Cowpea-88, Bundel lobia-1, Bundel lobia-2, DFC-1, CS-88, and UPC-622.

12.22 COWPEA HAY

In West Africa, cowpea hay is an important fodder sold in local markets. In smallholder systems, when used as a dual-purpose legume, cowpea hay can be used as animal feed. Cowpea can also be grown with Sudan grass for hay. When cowpea is grown specifically for hay, cowpea hay can be of similar quality to alfalfa hay. Hay yields are generally 3–5 t/ha. Hay quality declines as the crop matures (Cook et al., 2005). When cowpea is specifically grown for hay, cutting should be done when 25% of the pods are colored. In Australia, the ideal time to cut a cowpea crop for hay is at peak flowering, which occurs 70–90 days after sowing.

12.23 COWPEA SILAGE

Ensiling cowpea alone is not recommended as it is too moist. However, excellent silage can be made by harvesting a mixed crop of cowpea and forage sorghum, millet, or maize (Cook et al., 2005). The ability of cowpea to grow well in a maize crop and to climb the stems of maize makes harvesting both crops together possible (Teyneyck and Call, 1909). The intercropping of maize and cowpea at a seed ratio of 75:25; 70:30; or 50:50 has been reported in Iran, Pakistan, and Kansas to increase whole fodder production and to produce quality silage. Cowpea haulms (vines) can be used to make silage through the addition of water and 5% molasses. This ensiling process enhanced feed value but was not sufficient to fulfill the requirements of goats (Solaiman, 2007).

12.24 COWPEA PASTURE AND CUT-AND-CARRY SYSTEMS

Cowpea pastures and cut-and-carry systems are well developed in Asia and Australia. In Australia, cowpea forage is considered as annual forage whose quality is at his best during summer and autumn. When seasons are suitable, and when sown relatively early, the best forage types will regrow after grazing or cutting. In Kansas, cowpea was used for pasture in the early 20th century. It provided succulent feed during late summer when natural pastures were short. Used with maize, it was high-grade forage for pigs and sheep. Livestock had to be turned on cowpeas when the plant was fully developed (yellow pods) in order to prevent trampling and to provide its full feeding value. Cattle entered the swards before sheep and pigs, which allowed the latter animals to graze ripe cowpeas. The best feeding value was obtained in stands where cowpeas were intercropped with maize. In the drier parts of Kansas, it was suggested to intercrop cowpea with wheat or oats rather than maize (Teyneyck and Call, 1909).

Grazing should be light to ensure that most of the plant is preserved. Cowpea may suffer from trampling, if livestock enter the sward before the plants are full-grown. To prevent this, cowpeas can be used for zero grazing or can be grazed by pigs before cattle or sheep. Yields of fresh fodder can be increased by cutting the plants two or three times in a season. In Afghanistan, under irrigation, it is recommended to make the first cut after 60–65 days from sowing, the second cut 45–55 days later, and the third cut 50 days later (Oushy, 2012). When cut, cowpea can be mixed with dry cereals for stall-feeding.

12.25 CROPPING SYSTEMS

Intercropping of cowpea with sorghum/bajra/maize is quite prominent in South-central Deccan Plateau zone, north-western semiarid and arid zones, and central semi- and Vindhyan zone as the main source of fodder to the animals. In Himalayan foot hill regions, maize–cowpea–oat (fodder) cropping system is extensively followed. The most important sequential food cropping system involving cowpea is maize–potato–wheat–cowpea (fodder) in northern India and rice–rice–cowpea in Kerala. In coconut gardens of Kerala, cowpea, guinea grass, and *Stylosanthes guianensis* are grown. Further, in foothill regions of West Bengal and Assam, Dinanath grass (*Pennesetum pedicellatum*) + cowpea/rice bean/*Stylosanthes* intercropping systems are popular. Cowpea is also intercropped with guinea grass and Napier, bajra hybrids in Western and South India, respectively.

12.25.1 OVERLAPPING SYSTEM

The overlapping cropping system evolved by taking advantage of the growth periods of different species ensures a uniform supply of green fodder throughout the year, once such system runs for 3 years. The best rotation in this system is berseem + Chinese cabbage–hybrid Napier + cowpea–hybrid Napier (October–April)–(April–June)–(June–October), respectively.

12.26 USES

Cowpea is a quick-growing crop, which produces tremendous quantity of bulk mass in short span of time; and therefore, is esteemed as valuable catch crop or fodder, green manure, grains, or vegetable. The spreading indeterminate and semi determinate bushy growth of cowpea provides ground cover, thus suppressing the growth of weeds and providing protection against soil erosion by running water during heavy rains. A complete ground cover also reduces the temperature of the soil. As a legume, cowpea roots fix atmospheric nitrogen (about 40–80 kg/ha), thus increasing or improving the nitrogen content of the soil in which it is growing. It is of considerable importance in dryland farming and adapted to wide range of soils. In many parts of India, cowpea is grown as fodder crop, especially in summer and rainy season. In Kerala, fodder cowpea can be profitably grown as a summer crop in rice fallows, especially in rotation sequence, rice–rice–fodder cowpea and replenish the soil fertility for the next crop. The root, stem, and haulm residues decay after harvest, providing organic matter and the contained nutrients to the soil. All the parts of cowpea used for food (fresh leaves, immature pods, and the grains) are nutritious, providing protein, carbohydrate, vitamins, and minerals.

Well-cured cowpea haulms are useful feed and can make excellent hay, provided that the leaves are well preserved (too much exposure to the sun makes them fall off) and that the stems are adequately wilted. The haulm of fodder cowpea cultivars which could be used as cheap source of nitrogen supplement. Cowpea haulm has been shown to increase microbial nitrogen supply in calves when used as supplement to tuff straw (Abule et al., 1995), and to promote intake of maize stover and improve rumen ammonia concentration and degradation of maize stover (Chakeredza et al., 2002).

12.27 ANTINUTRITIONAL FACTORS

Cowpea contains some antinutritional factors which create some physiological disorders on consumption. Some of them are protease inhibitors, antivitamins, phytase, saponins, amylase inhibitors, tannins, aflatoxins, etc. Trypsin inhibitors which reduce protein digestibility can be destroyed and protein quality/digestibility can be improved on heating or autoclaving. However, simple heating and soaking may not easily remove heat-stable compounds (polyphenols and phytates). The same can be reduced on germination and/or fermentation. Polyphenols are found more in dark-colored seeds.

12.28 CONCLUSION

From the above discussion, it can be concluded that there are greater prospects for adoption and impact of dual-purpose fodder cowpea to alleviate poverty and malnutrition, and to contribute to the sustainability of tropical farming systems through profitable crop/livestock integration. Cultivation of fodder cowpea strengthens the household food security and supply of nutritious feed for the livestock, cash income, crop diversification, fodder, and in situ grazing after harvesting. However, to speedup dissemination and widespread adoption throughout the country, there is a need to involve national agricultural extension services and nongovernmental organizations as well as the private sector making seeds and planting materials available to a larger number of farmers, and also to provide preservation, processing, and marketing facilities on a participatory basis with the farmers' group for profitable production.

FIGURE 12.1 (See color insert.) Cowpea.

KEYWORDS

- **cowpea**
- **agronomic management**
- **fodder yield**
- **nutritive value**
- **antinutritional factors**
- **utilization**

REFERENCES

Abdel-Aziz, M. A.; Salem, M. F. Effect of Microbial Inoculation on Reduction of Cowpea (*Vigna unguiculata*, L. Walp) Chemical Fertilizers Under Newly Reclaimed Soils Condition in Egypt. *J. Plant Prod.* **2013,** *4* (5), 745–761.

Abule, E.; Umunna. N. N.; Nsahlai, I. V.; Osuji, P. O.; Alemu, Y. The Effect of Supplementing Tuff (*Eragrostis tef*) Straw with Graded Levels of Cowpea (*Vigna unguiculata*) and Lablab (*Lablab purpureus*) Hays on Degradation, Rumen Particle Passage and Intake by Crossbred (Friesian × Boran (zebu)) Calves. *Livest. Prod. Sci.* **1995,** *44*, 221–228.

Anita, M. R.; Lakshmi; S. Growth Characters of Fodder Cowpea Varieties as Influenced by Soil Moisture Stress Levels. *Indian J. Agric. Res*. **2015,** *49* (5), 464–467.

Anonymous. *Handbook of Agriculture*, 6th ed.; ICAR Publication: New Delhi, 2011.

Chakeredza, S.; ter Meulen, U.; Ndlovu, L. R.; Ruminal Fermentation Kinetics in Ewes Offered a Maize Stover Basal Diet Supplemented with Cowpea Hay, Groundnut Hay, Cotton Seed Meal or Maize Meal. *Trop. Anim. Health Prod.* **2002,** *34* (3), 215–223.

Choudhary, V. K.; Suresh Kumar, P. Nutrient Budgeting, Economics and Energetic of Cowpea Under Potassium and Phosphorus Management. *SAARC J. Agri.* **2013,** *11* (2), 129–140.

Cook, B. G.; Pengelly, B. C.; Brown, S. D.; Donnelly, J. L.; Eagles, D. A.; Franco, M. A.; Hanson, J.; Mullen, B. F.; Partridge, I. J.; Peters, M.; Schultze-Kraft, R. *Tropical Forages;* CSIRO, DPI & F (Qld), CIAT and ILRI: Brisbane, Australia, 2005.

Das, B.; Wagh, A. P.; Dod, V. N.; Nagre, P. K.; Bawkar, S. O. Effect of Integrated Nutrient Management on Cowpea. *TAJH* **2011,** *6* (2), 402–405.

Jha, A.; Razan Malla, K ; Sharma, M.; Panthi, J.; Lakhankar, T.; Krakauer, N. Y.; Pradhanang, S. M.; Dahal, P.; Shrestha, M. L. Impact of Irrigation Method on Water use Efficiency and Productivity of Fodder Crops in Nepal. *Climate* **2016,** *4*, 4. DOI: 10.3390/cli4010004.

Khalil, Z. M.; Salem, A. Kh.; Sultan, F. M. Water Stress Tolerance of Fodder Cowpea as Influenced by Various Added Levels of Potassium Sulphate. *J. Soil Sci. Agric. Eng.* **2015,** *6* (2), 213–231.

Madukwe, D. K.; Ogbuehi, H. C.; Onuh, M. O. Effects of Weed Control Methods on the Growth and Yield of Cowpea (*Vigna unguiculata* (L.) Walp.) Under Rain-fed Conditions of Owerri. *AEJAES.* **2012,** *12* (11), 1426–1430.

Mandal, M. K.; Pati, R.; Mukhopadhyay, D.; Majumdar, K. Maximising Yield of Cowpea Through Soil Test-based Nutrient Application in Terai Alluvial Soils. IPNI News, *Better Crops-India*. 2009, 28–30.

Muhammad, R. C.; Muhammad, J.;and Tahira, Z. M. Yield and Yield Components of Cowpea as Affected by Various Weed Control Methods Under Rain Fed Conditions of Pakistan. *Int. J. Agric. Biol.* **2003,** *9* (1), 120–124.

Ng, N. Q. Cowpea, *Vignia unguiculata* (Leguminosae papiliodeae). *Evolution of Crop Plants,* 2nd ed.; Smatt, J, Simmonds, N. W., Eds.; Longman: New York, 1995; pp 326–332.

Nnadi, L. A.; Haque, I. Forage Legumes in African Crop–Livestock Production Systems. *ILCA Bull.* **1988,** *30*, 10.

Oushy, H. Factsheet: Forage Cowpea. New Mexico State University, USAID-Afghanistan Water, Agriculture, and Technology Transfer (AWATT) Program. 2012.

Parasuraman, P. Weed Management in Rain Fed Cowpea (*Vigna unguiculata*) and Greengram (*Phaseolus radiatus*) Under North West Agro Climatic Zone of Tamil Nadu. *Indian J. Agron.* **2000,** *45*, 732–736.

Patel, B. N.; Jadav, D. K. Effect of *Rhizobium* Seed Inoculation, Nitrogen and Phosphorus on Growth, Nodulation, Flowering and Seed Yield of Cowpea cv. Pusa Phalguni (*Vigna unguiculata* Walp). *Int. J. Agric. Sci.* **2010,** *6* (2), 361–364.

Solaiman, S. Feeding Value of Seed-harvested Cowpea Vines for Goats. Tuskegee University, Notes on Goats, Technical Paper. 2007, 7–9.

Teyneyck, A. M.; Call, L. E. Cowpeas. Kansas State Agricultural College, Experiment Station, Bulletin. 1909, 60.

CHAPTER 13

CLUSTER BEAN (GUAR)

SAVITRI SHARMA*

Department of Agriculture, Jagannath University, Chaksu 303901, Jaipur, India

**E-mail: savisharma423@gmail.com*

ABSTRACT

It is annual arid and semiarid legume crop grown during *kharif* season in India. The term guar has been evolved from the common use of the crop and its residue as cattle feed. Guar plant produces a cluster of flower and pods therefore, it is also known as cluster bean. Cluster bean is cultivated for its green vegetables and dry pods and as forage crop, and also cultivated for green manure because guar planting increase subsequent crop yields, as this legume crop conserves soil nutrients. A good crop under favorable climatic conditions yields about 300–400 quintals of green fodder or 15 quintals of dry seeds or 60 quintals of green pods per hectare.

13.1 BOTANICAL CLASSIFICATION

Kingdom: Plantae
Orders: Fabales
Family: Fabaceae
Subfamily: Faboideae
Genus: *Cyamopsis*
Species: *tetragonoloba* L.
Binomial name: *Cyamopsis tetragonoloba* (L.)

13.2 COMMON NAMES

Guar/gwar (Hindi); cluster bean (English); *goruchikkudu kaya* or *gokara-kaya* (Telugu); *chavalikayi* (Kannada); and *kothavarai* (Tamil).

13.3 BOTANICAL NAME

Cyamopsis tetragonoloba (L.)

13.4 INTRODUCTION

Guar is an important cash crop in rainfed, especially in semiarid and arid regions of India. It is annual arid and semiarid legume crop grown during *kharif* season in India. The term guar has been evolved from the common use of the crop and its residue as cattle feed. Guar plant produces a cluster of flower and pods therefore, it is also known as cluster bean.

Cluster bean is cultivated for its green vegetables and dry pods and as forage crop, and also cultivated for green manure because guar planting increase subsequent crop yields, as this legume crop conserves soil nutrients. Cluster bean occupies an important place in the national economy because of its industrial importance, mainly due to presence of gum in its endosperm. It has been estimated that on the whole-seed basis, guar seed contains about 30.7% gum. This gum is used in paper industry, explosives, hydrofracturing while drilling for shale gas, mining, and various food products such as ice cream, cheese, salad dressings, fruit drinks, and pharmaceuticals, etc. In all, there are more than 30 applications.

13.5 COMPOSITION

The guar seeds are dicotyledonous containing two endosperm halves per seed. The endosperm accounts for about one-third of the bean weight and contains the majority of wonderful galactomannan (gum). The remaining two-thirds are the hull and germ, which are very high in protein and fiber. Purity of the polymer depends on effective separation of the hull and germ from the endosperm. Endosperm halves (splits) may be processed or modified and compounded into guar products with specialized properties. Approximately 90% of total guar produce is used for production of commercial guar gum and rest is used for culinary purposes, cattle feed, etc. Guar seed endosperm contains protein 5%, ether extract 0.6%, ash 0.6%, moisture 10%, and fibers 1.5%.

13.6 IMPORTANCE AND UTILITY

Cluster bean or guar among leguminous crops is comparatively more drought-hardy crop. Cluster bean is usually grown for feed, fodder, and vegetable purposes. Among dry land crops, cluster bean occupies an important place in the national economy because of its industrial importance, mainly due to presence of gum in its endosperm. Its gum is used in various industries. The fodder of cluster bean as well as its grain is quite nutritive, rich in protein, fat, and minerals. It can also be used as green manure crop in certain areas.

13.7 DESCRIPTION

Cluster bean belongs to the family Leguminosae. Its plant is erect, robust, and annual, and grows to a height of 90–180 cm. Some varieties may grow even taller than this. Its plant has well developed taproot system. It has a main single stem with either basal branching or fine branching along the stem. The leaves are trifoliate and toothed. The flowers are borne in short axillary racemes and generally purplish in color. The developing pods are rather flat and slim containing 5–12 small oval seeds of 5 mm length. Usually, mature seeds are white or gray, but in case of excess moisture they can turn black and lose germination capacity. When pods are tender they are used as vegetable.

13.8 ORIGIN AND DISTRIBUTION

Cluster bean is being grown in India since ancient time for vegetable and fodder purposes. Therefore, some people think that guar is probably indigenous to India. Gillete (1958) pointed out that tropical Africa is the probable center of origin because of more occurrences of wild species in that country. The cultivation is largely confined to North India, comprising Rajasthan, Haryana, Punjab, and Uttar Pradesh. It is also grown in Gujarat and Madhya Pradesh. The total area under its cultivation is 0.2 million hectare.

13.9 CLIMATIC REQUIREMENT

Guar is a photosensitive crop. It grows in specific climatic condition, which ensure a soil temperature around 25°C for proper germination, long photoperiod with humid air during its growth period, and finally short photoperiod with cool dry air at flowering and pod formation. Accordingly, it is definitely a *kharif* season crop in North India, but some varieties have been found to grow during March to June as spring–summer crop and other varieties grow during July to November as rainy season crop under South Indian climatic conditions. It is a crop preferring warm climate and grows well in the subtropics during summer. Guar tolerates high temperatures and dry conditions and is adapted to arid and semiarid climates. Optimum temperature for root development is 77–95°F. When moisture is limited, the plant stops growing but does not die. While intermittent growth helps the plant survive drought, it also delays maturity. Growing season ranges from 60–90 days (determinate varieties) to 120–150 days. Guar responds to irrigation during dry periods. It is grown without irrigation in areas with 10–40 in. of annual rainfall. Excessive rain or humidity after maturity causes the beans to turn black and shrivel, reducing their quality and marketability (Singh et al., 2009).

13.10 SOIL AND ITS PREPARATION

Guar grows well under a wide range of soil conditions. It performs best on fertile, medium-textured, and sandy loam soils with good structure and well-drained subsoils with pH range of 7.0–8.5. Guar is considered to be tolerant to both soil salinity and alkalinity, but is susceptible to waterlogging.

Guar crop requires a well prepared field, with adequate soil moisture for its seed germination. During early growth period, soil aeration encourages root development and bacterial growth. Therefore, field should be plowed to fine tilth by giving two or three deep plowing with soil-turning plow, followed by harrowing and planking. The field should be free from weeds and other crop residues. Arrangements for drainage channel-cum-water channel for heavy rainfall areas or irrigated areas should be made while preparing the field. The farmyard manure should be mixed with soil at the time of last plowing.

13.11 VARIETIES

13.11.1 DURGAJAY

The variety was developed at Agriculture Research Station (ARS), Durgapura from single plant selection of the material collected from Nagaur, Rajasthan. The variety is dual type for fodder and seed and is recommended for cultivation in Rajasthan. The variety provides 27 t/ha green fodder and seed yield is 12.6 quintals/ha.

13.11.2 DURGAPURA SAFED

This variety was developed by ARS, Durgapura by single plant selection from local material of Rajasthan. It is a dual-type variety suitable for late sown conditions. It is recommended for cultivation in Rajasthan state. It yields about 14–15 quintals/ha and green fodder yield is 25 t/ha.

13.11.3 AGETA GUARA-111

The variety was developed by Punjab Agriculture University (PAU), Ludhiana from intervarietal cross of G 325 and FS 277 followed by pedigree method of selection. The variety is recommended for cultivation in all guar-growing areas of Punjab state. It is an early maturing variety which takes about 95–110 days to mature. It provides 23 t/ha of green fodder and 4.4 t/ha dry fodder.

13.11.4 BUNDEL GUAR-3 (IGFRI 1019-1)

This variety was developed by Indian Grass and Fodder Research Institute (IGFRI), Jhansi through selection from indigenous material collected from Durgapura, Rajasthan (RGC-19-1). The variety is moderately resistant to bacterial blight and powdery mildew, responsive to fertilizers, highly tolerant to shattering, and reasonably resistant to drought situations. The variety has been released and notified for general cultivation in entire guar-growing area of India as forage-cum-grain type. The maturity is 50–55 days producing 35–40 t/ha green fodder.

13.11.5 HG-20: RELEASED IN 2008

This variety is recommended for Punjab, Haryana, Gujarat, Rajasthan, and Uttar Pradesh under timely sowing, normal fertility, and rainfed condition. The plants mature in about 90–100 days. To get maximum production, sowing may be done during last week of June to first week of July at 45 cm row spacing. The plant height is about 70 cm and it has pubescent and serrated leaves. This variety has field tolerance to bacterial blight, moderately resistance to Alternaria blight and root rot.

13.11.6 PUSA SADABAHAR

This variety is suitable for green pod production. It is photo insensitive and grows for longer duration producing green pods for pretty sufficient period. It can be sown as early as March and starts producing pods after 45 days of sowing. Plucking of pods continues till September. It is also suitable for green fodder and green manuring. The disadvantage with this variety is poor quality of pods. It yields about 60–70 quintals of green pods per hectare. It is recommended for South India as rainfed crop.

13.11.7 PUSA MAUSAMI

It is recommended for vegetable pod production in Punjab, Haryana, Delhi, and Uttar Pradesh. Its plants are unbranched; produce slender, attractive, smooth pods of 1.2-cm length. The pods being nonfibrous are used as vegetable. The first plucking of pods begins 80 days after sowing. This variety can be grown under rainfed as well as irrigated conditions. It yields about 50 quintals of green pods per hectare.

13.11.8 RGC-1066 (GUAR LATHI)

It was released in 2007. It can be grown under rainfed and well-drained soil. Its plants mature in about 97–105 days. Plant height is about 86–110 cm, medium tall 86–110 cm, and its plant has unbranched leaves. Grain weight is 3.38 g/100 seeds. It is moderately resistant to bacterial and Alternaria blight, root rot, and wilt.

13.11.9 RGC-1055 (GUAR UDAY)

This variety is recommended for all the states of the northern India. It can be grown under rainfed conditions. It matures in about 96–106 days. Plants are of medium height, 74–97 cm, and are profusely branched. Grain weight is 3.38 g/100 seeds. It is moderately resistant to bacterial and *Alternaria* blight, root rot, and wilt.

13.12 SOWING TIME

Guar is a *kharif* season crop in North India, but some varieties have been found to grow during March to June as spring–summer crop and other varieties grow during July to November as rainy season crop under South Indian climatic conditions. It is a crop preferring warm climate and grows well in the subtropics during summer. Summer crop of guar in northern

India is sown in March, while *kharif* crop in June. For grain crop, the best time of sowing is during July. Early sowings result in more vegetative growth, lodging, and loss of yield. The best sowing time for fodder crop is April. In peninsular India, guar is sown in September. In South India, it is sown at any time between February and October. Guar sowing time is adjusted to May–June to increase supply of vegetable pods. Crop sown in June in Haryana and in July in Punjab gives better production of vegetable pods.

13.13 SEED TREATMENT

In humid areas, where there are more chances of disease outbreak, seeds are treated in the following ways:

- Dry seed is coated with Ceresan or Thiram at the rate of 3 g fungicide per kilogram of seed to kill the spores of fungus resting on the seed coat.
- Seed is immersed in hot water at 56°C for 10 min and then dried at room temperature before sowing. This kills all the fungus mycelium and inactivates their spores which spread disease in the crop.

13.14 SEED RATE AND SOWING METHOD

Seed rate of guar crop varies from 15–40 kg per hectare depending upon soil moisture, spacing, and purpose of cultivation. About 15–25 kg seed per hectare is sufficient for grain crop. Seed rate for crops grown for fodder or green manuring is about 40 kg per hectare. Seed rate is normally increased under late sown condition, dry condition, and soil salinity or alkalinity conditions.

Seed of guar can be sown by broadcast method, when there is sufficient soil moisture during the beginning of the rainy season. Line sowing with the help of *pora* behind the plow or by seed drill is useful in sowing the seed at proper spacing and depth. This results in better seed germination. Line sowing is also useful for carrying out hoeing–weeding and removing excessive rain water.

13.15 SPACING

A spacing of 45 cm between rows and 15 cm between plants is given for crops grown for grain production. However, spacing is generally reduced under late sowing and poor soil fertility conditions. Closer spacing of 30 cm × 12 cm is provided for fodder crop. A wider spacing of 60 cm × 30 cm is desirable for crops grown for green vegetable pod production.

13.16 CROPPING SYSTEM

Cluster bean is usually grown in rotation with *rabi* crops. For fodder purpose, following rotations are adopted in northern India.

Cluster bean–berseem–maize + cowpea

Pearl millet + cluster bean–berseem

Sorghum + cluster bean–berseem–maize + cowpea

13.17 NUTRIENT MANAGEMENT

Guar crop needs 10–12 t of well decomposed farmyard manure, especially when it is being cultivated on poor sandy soils, or after taking an exhausting crop. This manure is applied a month before sowing. Nitrogenous fertilizers are applied only in small quantity (about 20 kg of nitrogen) because most of nitrogen to the crop comes from the atmosphere through bacterial action. Phosphatic fertilizers about 60 kg, potassic 20 kg per hectare along with 20 kg of nitrogen are applied as basal dose at the time of sowing, with the help of *pora* just 4–5 cm below the seed. Spray of crop with 0.15% solution of sodium molybdate after 30 days of sowing increases the yield of both fodder and grain.

13.18 WATER MANAGEMENT

The *kharif* season crop grown during rainy season as a rule does not need any irrigation if rains are adequate and well distributed. If rains are too heavy, the excess water should be drained off promptly. For dry season crops, irrigation at fortnightly intervals in the early summer and at 10 days intervals later is given. Normally the crop requires 4–5 irrigations.

13.19 WEED MANAGEMENT

Guar field in *kharif* season is always full with a number of weed plants, hoeing and weeding in the initial stages of plant growth with the help of *khurpi*; tined harrow reduces the weed–crop competition and increases soil aeration for bacterial growth. The application of Basalin at the rate of 1.0 kg a.i. per hectare as preplanting dose suppresses the growth of grasses and other weeds.

13.20 PLANT PROTECTION MEASURES

13.20.1 DISEASES

The most common diseases of guar are wilt, bacterial blight, powdery mildew, and anthracnose. These diseases along with their control measures are described below.

13.20.1.1 WILT

This disease is caused by *Fusarium moniliformae*. The pathogen is soil-borne and, therefore, its damage is of localized nature. It causes infection to the base of plants including roots and seedlings. The roots show discoloration and plant wilts. Poor emergence of seedlings is the first symptom of the disease. The seedlings rot before or soon after emergence.

Control

- Mixed cropping of guar with sorghum reduces the damage up to 55%.
- Addition of organic manures also reduces the disease incidence.
- Seed treatment with Agrosan GN or Thiram or captan at the rate 3 g/kg seed prevents the disease spread during germination and seedling emergence.

13.20.1.2 ANTHARACNOSE

This disease is caused by *Colletotrichum capsici*. The disease is more severe in high rainfall subtropical to temperate areas than in tropical areas. The fungus is seed borne and symptoms may start as early as in seedling stage. The most characteristic symptoms of the disease are black, sunken, crater-like cankers on the pods, stem, or cotyledons. The lesions remain isolated by yellow-orange margins.

Control

- Use healthy seed.
- Avoid excess watering.
- Give wider spacing.

- Use hot water treated seed.
- Spray fungicides such as Dithane M-45 or Dithane Z-78 at the rate of 2 kg in 1000 L of water per hectare.

13.20.1.3 POWDERY MILDEW

This disease is caused by *Oidium* sp. White powdery growth occurs on leaves, spreading to cover the stem and other plant parts. In severe cases, the entire plant dries up.

Control

- Dusting with sulfur powder.
- Spraying with systemic fungicides Benlate or Bavistin and Calixin gives effective control of powdery mildew.
- Sowing healthy seed after treating with Thiram.
- Follow a crop rotation to reduce the soil-borne inoculum of the fungus.

13.20.1.4 BACTERIAL BLIGHT

This disease is caused by *Xanthomonas cyamophagus*. The disease is characterized by irregular, sunken, red to brown leaf spots surrounded by a narrow yellowish halo. Several spots coalesce to form irregular patches. The spots may also develop on pods.

Control

- Grow resistant varieties.
- Treat the seed with Thiram at the rate of 3 g/kg seed. Also adopt hot water treatment at 56°C for 10 min.
- Eradicate affected plants and burn them.

13.20.1.5 LEAF SPOT

This disease is caused by *Myrothecium roridum*. Dark brown round spots appear on leaf. In case of severe infection, several spots merge together and leaflets become chlorotic and usually drop off. If plants are infected in the early stages of growth, there may not be any flowering.

Control

Spray with Dithane Z-78, 0.2% at the interval of 15 days, twice or thrice.

13.20.2 INSECT CONTROL

Guar is a rainy season crop. Therefore, a number of insect-pests feed and grow on its leaves and pods. Some important ones are described here.

13.20.2.1 HAIRY CATERPILLARS

Ascotis imparata and *Spilosoma obliqua, Amsacta lactinea* and *Euproctis scintillans* are the insects that cause heavy damage. The adult lays eggs in clusters over the surface of the leaves. The larvae cause characteristic skeletonization of leaves during the early gregarious stage and later they completely defoliate the plant. The pest can easily be controlled by systematic collection of larvae during the early gregarious stage.

13.20.2.2 JASSID

Empoasca fabae, *Empoasca kraemeri*, and *Amrasca kerri* are serious pests of the crop. The nymphs are wingless and found in abundance on the lower surface of the leaves. The nymph and adult pierce the plant tissues and suck the cell sap. Leaves become yellow at the margin. They can be controlled by:

1. Soil application of systemic granular insecticides such as Aldicarb 10 G at the rate of 10–15 kg per hectare.
2. Spray the crop with 0.04% monocrotophos or 0.04% metasystox.

13.21 HARVESTING AND YIELD

Green pods are used as vegetable purpose. The pods of the guar become ready for plucking, depending upon the variety, from 40 days onwards after sowing. Picking is done at an interval of 10–12 days. When crop is grown for fodder, the plants are cut when they are at flowering stage or when the pods are beginning to emerge. This stage comes 50–80 days after sowing. For green manuring, the crop can be plowed down as soon as the pods begin to develop. The yield of the green material crop is about 120 quintals per hectare. When guar is grown for seed, the crop is retained in the field until the pods are mature. A good crop under favorable climatic conditions yields about 300–400 quintals of green fodder or 15 quintals of dry seeds or 60 quintals of green pods per hectare.

KEYWORDS

- **guar**
- **agronomic management**
- **fodder yield**
- **nutritive value**
- **utilization**

REFERENCES

Gillete, J. B. Indigofera (Microcharis) in Tropical Africa with the Related Genera Cyanopsis and Rhynchoptnopsis. *Kew Bull. Add. Serv.* **1958,** *1*, 1–166.

Singh, C.; Singh, P.; Singh, R. *Modern Techniques of Raising Field Crops*; Oxford & IBH Publishing: India, 2009.

FIGURE 1.1 Oats fodder: before harvesting (Var.—OS-6).

FIGURE 2.1 Fodder maize.

FIGURE 3.1 Pearl millet.

FIGURE 4.1 Sorghum.

FIGURE 6.1 Coix.

FIGURE 10.1 Hybrid Napier.

FIGURE 11.1 Guinea grass.

FIGURE 12.1 Cowpea.

FIGURE 14.1 Rice Bean.

FIGURE 15.1 Berseem.

FIGURE 16.1 Lucerne.

FIGURE 21.1 Stack silo.

FIGURE 21.2 Bunker silo.

FIGURE 21.3 Plastic bag silo.

FIGURE 21.4 Tower silo.

CHAPTER 14

RICE BEAN (RED BEAN)

CHAMPAK KUMAR KUNDU*

*Department of Agronomy, Faculty of Agriculture,
Bidhan Chandra Krishi Viswavidyalaya, Mohanpur 741252,
Nadia, West Bengal, India*

**E-mail: champakbckv@gmail.com*

ABSTRACT

Rice bean is a tropical to temperate annual vine grain legume with yellow flowers and small edible beans. It is regarded as a minor food, especially in Asia, and rice bean seeds and vegetative parts are also used for fodder. Rice bean grows well on a range of soils. It establishes rapidly and has the potential to produce large amounts of nutritious animal fodder and high-quality grain. Green forage yield may be obtained 175–350 quintals/ha from summer crop and 200–350 quintals/ha from kharif crop. Rice bean seeds contain antinutritional factors. Antitrypsin activity is notable though comparable to that of cowpea. Rice bean contains phytic acid, polyphenol, tannins, trypsin inhibitors, other antinutrients, and flatus-producing oligosaccharides. Rice bean is nitrogen-fixing legume that improves the N status of the soil, thus providing N to the following crop and enhances the fertility of soil to a great extent.

14.1 BOTANICAL CLASSIFICATION

Kingdom: Plantae
Order: Fabales
Family: Fabaceae
Subfamily: Faboideae
Tribe: Phaseoleae

Genus: *Vigna*
Species: *umbellate*
Binomial name: *Vigna umballate* (L.)

14.2 SCIENTIFIC NAME

Vigna umballate (L.)

14.3 COMMON NAMES

Rice bean, red bean, rice bean, climbing mountain bean, mambi bean, oriental bean (English); haricot. It is popularly known as rice bean. Its common name in Bengal is *Gai mung* or *Gai Kalai*, in Tamil, *Karum payaru*, and in Telegu, *Karu pesara* and Sutri.

14.4 SYNONYMS

Azukia umbellata (Thunb.) Ohwi, *Dolichos umbellatus* Thunb. *Phaseolus calcaratus* Roxb., and *Vigna calcarata* (Roxb.) Kurz. The sysnonyms are *Phaseolus sublobatus* Walt. *P. hirtus* wall, *P. receiardianus* Ten., *P. torosus* Roxb, *P. calcaratus* Roxb, and *Azukia umbellata* (Thunb.) Ohwi.

14.5 TAXONOMY

Rice bean is an annual food legume belonging to the subgenus *Ceratotropis* in the genus *Vigna*. The genus *Vigna*, together with the closely related genus *Phaseolus*, forms a complex taxonomic group, called *Phaseolus–Vigna* complex. Verdcourt (1970) proposed a very restricted concept of *Phaseolus*, limiting it exclusively to those American species with a tightly coiled style and pollen grains lacking course reticulation, hence, promoting significantly the concept of *Vigna*. According to his proposal, rice bean and its relatives were transferred to the genus *Vigna* from the genus *Phaseolus*. Marechal et al. (1978) followed Verdcourt and presented a monograph on the *Phaseolus–Vigna* complex. Their taxonomic system is generally accepted now.

In their monograph, two varieties are recognized in *Vigna umbellata*. One is *V. umbellata* var. *umbellata* which is rice bean, and another is *V. umbellata* var. *gracilis* which is a wild ancestor of rice bean. On the other hand, Tateishi (1985) considered that the specimen of *V. umbellata* var. *gracilis* described in the monograph of Marechal et al. (1978) should be included in *Vigna minima*. 2n = 22. The cultivated Asiatic *Vigna* species belong to the subgenus Ceratotropis, a fairly distinct and homogeneous group, largely restricted to Asia, which has a chromosome number of 2n = 22 (except *Vigna glabrescens*, 2n = 44). There are seven cultivated species within the subgenus, including mung bean or green gram (*Vigna radiata*), black gram or urad bean (*Vigna mungo*), adzuki bean (*Vigna angularis*), and moth bean (*Vigna aconitifolia*) as well as a number of wild species. Artificial crosses have been made between *V. mungo* and *V. umbellata* to produce improved mung bean varieties (Singh et al., 2006).

14.6 INTRODUCTION

V. umbellata (Thunb.) Ohwi and Ohashi, previously *P. calcaratus*, is a tropical to temperate annual vine grain legume with yellow flowers and small edible beans. It is regarded as a minor food, especially in Asia, and rice bean seeds and vegetative parts are also used for fodder and is often grown as intercrop or mixed crop with maize (*Zea mays*), sorghum (*Sorghum bicolor*), or cowpea (*Vigna unguiculata*), as well as a sole crop in the uplands, on a very limited area.

Rice bean is a fairly short-lived warm-season annual. Grown mainly as a dried pulse, it is also important as a fodder, a green manure, and a vegetable. In the past, it was widely grown as lowland crop on residual soil water after the harvest of long-season rice, but it has been displaced to a great extent where shorter duration rice varieties are grown. Rice bean grows well on a range of soils. It establishes rapidly and has the potential to produce large amounts of nutritious animal fodder and high-quality grain.

14.7 ORIGIN AND HISTORY

Rice bean originated from Indochina and was probably domesticated in Thailand and neighboring regions. Rice bean is a native of Southeast Asia and occurs in India in wild and cultivated forms. Wild forms of rice bean are distributed in northeastern India, Burma, Thailand, Laos, and Vietnam. Its grains resemble dehusked rice grains, hence the name rice bean. However, the trifoliate leaves and stems resemble that of green gram.

It is also a grain legume grown in hilly areas of Nepal and northern India. It is an underutilized crop which is grown by only resources lacking farmers. It has the potential to play a major role in intercrop systems, and can also be grown on the edges of the terraces or sloping hillsides, as well as on rice bunds and under shifting cultivation on hillsides. The popularity of rice bean as legume forage is increasing gradually because of its wider adaptability, high productions, and surviving capacity through winter. Its potential to be a boon in improving the livelihoods of many poor people is yet to be fully tapped. Today when the world is booming with scientific advancements, chances are full that simple steps will make this crop a huge range as a widely cultivated crop.

14.8 GEOGRAPHICAL DISTRIBUTION

Rice bean is a tropical legume and is native to South and Southeast Asia. In India, it occurred both as wild and cultivated types. It is cultivated in West Bengal, Assam, Madhya Pradesh, and Haryana in India. Other countries where rice bean is grown are Burma, Indonesia, Uganda, Paraguay, Columbia, Malagasy Republic, Brazil, and the southern states of the United States. But in some countries root knot nematodes restrict its cultivation.

Rice bean distribution pattern indicates great adaptive polymorphism for diverse environments, with its distribution ranging from humid tropical to subtropical to subtemperate climate. It is found naturally in India, central China, and in the Indochinese Peninsula. It was introduced to Egypt, to the east coast of Africa, and to the islands of the Indian Ocean. It is now cultivated in tropical Asia, Fiji, Australia, tropical Africa, the Indian Ocean islands as well as in the Americas (Rajerison, 2006; Khadka et al., 2009). In the middle hills of Nepal, rice bean is cultivated along rice bunds and terrace margins (Khadka et al., 2009). Though it can thrive in the same conditions as cowpea and can better tolerate harsh conditions (including drought, waterlogging, and acid soils), rice bean remains an underutilized legume and there is no breeding program to improve this crop. Farmers must rely on landraces rather than on cultivars (Joshi et al., 2008).

Rice bean is a fast summer-growing legume found from sea level up to altitudes of 1500 m in Assam and 2000 m in the hills of the Himalayas (Khadka et al., 2009). Rice bean requires a short day length to produce seeds. It is grown on a wide range of soils, including shallow, infertile, or degraded soils. High soil fertility may hinder pod formation and reduce seed yield (Khadka et al., 2009). *V. umbellata* is a versatile legume that can grow in humid subtropical to warm and cool temperate climates. It is suited to areas with annual rainfall ranging from 1000 to 1500 mm but it is also fairly tolerant of drought. It does better in areas where average temperatures range from 18°C to 30°C, tolerates 10–40°C but does not withstand frost (Rajerison, 2006). It prefers full light and its growth can be hampered if it is intercropped with a tall companion crop that overshadows it, such as, maize (Khadka et al., 2009).

Rice bean is a tropical legume and is native to south and Southeast Asia. In India, it is cultivated in West Bengal, Assam, Madhya Pradesh, and Haryana. Other countries where rice bean is grown are Myanmar, Indonesia, Uganda, Paraguay, Columbia, Malagasy Republic, Brazil, and southern states of the United States.

According to the literature, rice bean was grown in the southwestern part of Japan formerly. It was called "Me-naga (long eye)" in Tsushima Island and "Sasage or Kaninome" in Goto Island in Nagasaki prefecture, "Kani-no-me (eye of a crab)" in Yamaguchi prefecture, "Baka-azuki (foolish azuki bean)" in Kagoshima prefecture, and "Kage-azuki (shady azuki bean)" in Tottori prefecture. Recently, cultivation of this crop has nearly disappeared in Japan.

14.9 ADAPTION

Rice bean adapted to subhumid regions with 1000–1500 mm precipitation, although they noted that other factors were also involved in adaptation, for example, rainfall pattern, moisture distribution, temperature, cloud cover and relative humidity, soil characteristics, pests, and diseases. Average yields were between 200 and 300 kg/ha, although with the potential for 1200 kg/ha, the crop would grow on a range of soils, and was resistant to pests and diseases. It would mature in as little as 60 days, and although performing well under humid conditions, was also tolerant to drought and high temperatures. It is tolerant to some degree of waterlogging, although the young plants appear to be susceptible. Rice bean is also known to be tolerant to acid soils. Shattering is a problem in comparison with other grain legumes, and can be particularly serious under conditions of frequent wetting and drying.

Rice bean is a neglected crop, cultivated on small areas by subsistence farmers in hill areas of Nepal, northern and northeastern India, and parts of Southeast Asia. It can be grown in diverse conditions and is well known among farmers for its wide adaptation and production even in marginal lands, drought-prone sloping areas, and flat rainfed *tars* (unirrigated, ancient alluvial river fans). It is mainly grown between 700 and 1300 m also, although in home gardens it is found from 200 up to 2000 m. Most of the crop currently grown in Nepal is used as food for humans, with a smaller proportion used for fodder and green manuring.

Generally, rice bean is grown as an intercrop with maize, on rice bunds or on the terrace risers, as a sole crop on the uplands, or as a mixed crop with maize in the *khet* land. Under mixed cropping with maize, it is usually broadcast sometime between sowing maize and that crop's first and second earthing up, so rice bean sowing extends from April–May to June.

Rice bean is valuable for its ability to fix nitrogen in depleted soils and in mixed cropping with local varieties of maize, as well as for its beneficial role in preventing soil erosion. The crop receives almost no inputs, and is grown on residual fertility and moisture and in marginal and exhausted soils. Anecdotal evidence indicates that the area and production of rice bean in Nepal is declining due to the introduction of high-yielding maize varieties and increasing use of chemical fertilizers, while consumption is decreasing due to increased availability of more preferred pulses in the local markets. No modern plant breeding has been done and only landraces with low yield potential are grown. These have to compete with other summer legumes such as soybeans (*Glycine max*), black gram, cowpea, common beans (*Phaseolus vulgaris*), and horse gram (*Macrotyloma uniflorum*). Other production constraints that limit the production of rice bean include small and fragmented land holdings and declining productivity.

Rice bean is most widely grown as an intercrop, particularly with maize, throughout Indochina and extending into southern China in the east and into northeast India, Bangladesh, and Nepal in the west. It is grown mainly as a dried pulse, but also used as a fodder crop and a source of green pods.

In the past, it was widely grown as a lowland crop on residual soil water after the harvest of traditional long-season rice varieties. However, it has been substantially displaced where multiple cropping of shorter duration rice varieties has expanded. Despite their long history, only mung bean has received in-depth research interest. The mung bean, black gram, and adzuki bean are used in mechanized agriculture due to their more upright habit and relatively synchronous flowering.

14.10 PLANT MORPHOLOGY

V. umbellata is a short-lived perennial legume usually grown as an annual. It has a very variable habit: it can be erect, semierect, or viny twining. It is usually 30–100 cm in height, but can grow up to 200 cm (Ecoport, 2014). It has an extensive root system with a taproot that can go as deep as 100–150 cm. The wild forms are typically fine-stemmed, freely branching profusely and finely haired, photoperiod sensitivity, and indeterminate growth. The leaves are small, pinnately trifoliate with entire, 6–9-cm long leaflets. The flowers are asynchronous, borne in clusters and self-fertile but some natural outcrossing even interspecific occurs, born on 5–10-cm long axillary. Inflorescence is raceme, and is papilionaceous and bright yellow. The fruits are cylindrical, 7.5–12.5-cm long pods that contain 8–12 oblong/slender, 6–8-mm seeds with a concave/protruding hilum. Pod attaches to the peduncle downward. Germination is hypogeal. The plant is very hairy and the seed pods twist and shatter the seeds on ripening. The plant is photoperiod sensitive and the reproductive phase covers mid-November to February. Rice bean seeds have a tendency toward hard seeds, are very variable in color, from greenish-yellow to black through yellow-brown. Yellow-brownish types are reported to be the most nutritious. In many areas, landraces which retain many of these characteristics persist, in particular with regard to daylight sensitivity, growth habit, and hard seeds.

14.11 ENVIRONMENTAL REQUIREMENTS

The crop requires a warm humid tropical climate. It is fairly resistant to drought but susceptible to water logging, particularly in the early stage. Conducive temperature being 25–30°C, it grows only in tropical climate.

It grows in high rainfall areas where excess water is freely drained out. In fact, its climatic requirements are similar to those of cowpea, velvet bean, green gram, maize, *jowar*, etc. Sandy loam to clayey loam soils having sufficient calcium and phosphorus are suitable. The important thing is the land should be well fertilized and the plots should be well drained to avoid water stagnation.

14.12 LAND PREPARATION

Three to four cultivations with *desi* plow or three to four operations by a cultivator or a disc harrow with a tractor are necessary to bring the land for sowing. The fields should be cleaned of weeds as the delicate young plants are likely to be smothered in the early stage by the robust weeds.

14.13 VARIETY

Bidhan Ricebean-1, Bidhan Ricebean-2, Bidhan Ricebean-3, RBL-6, RBL-36, and Tagore Rain.

14.14 SOWING TIME

The rainfed crop may be sown any time between June and September. With irrigation facilities, sowings can be taken up from February onwards.

14.15 SOWING METHOD

The seeds are usually broadcast and covered with disc harrow, cultivator, or *desi* plow. The seeds may be sown in lines by drill, *pora*, or *kera* methods for better germination, uniform stand, and economy of seed.

14.16 SPACING

The spacing should be 30–40 cm between lines and 10–15 cm between plants.

14.17 SEED RATE

For forage, 40–45 kg of seed/ha is required in broadcast sowing and 30–35 kg/ha in line sown crop. When seeds are sown in standing paddy crop, 75 kg of seed/ha are used as some seeds fail to germinate in this method of sowing. When rice bean is sown in mixtures with cereal fodders, the seed rate is reduced to half. For grain purpose, about 30–45 kg/ha is sufficient.

14.18 MANURES AND FERTILIZERS

The field may be manured at the rate of 7.5–10 t/ha of farmyard manure at the time of land preparation and 40–60 kg P_2O_5 and 20–40 kg K_2O along with 20 kg N per hectare may also be applied just before the final land preparation. No top dressing is generally recommended. But, when first cut is taken for green fodder purpose and rest for seed or grain use, then top dressing is advocated with 20 kg N and 30 kg K_2O with irrigation.

14.19 IRRIGATION

If necessary, one presowing irrigation is important. Subsequent irrigations may be required every three to fourth week, if rains are not received. During the rainy season, drainage of water is essential. In the postmonsoon period, the crop may require one irrigation before it is finally harvested. The land should be kept well drained.

14.20 WEED CONTROL

In the usual course, rice bean requires clean cultivation and one or two weedings or interculture operations for establishment of the crop. The grassy and broad-leaved weeds which are prevalent in rice bean plots can be controlled chemically by herbicides applied as preemergence. These herbicides are S-ethyl dipropyl thio carbamate at the rate of 3–4 kg/ha (preemergence) or trifluralin/nitralin at the rate of 1.0–1.5 kg/ha applied 7 days before planting.

14.21 CROP MIXTURES

In mixed cropping, rice bean can be sown with maize, *jowar, bajra*, teosinte, Sudan grass, and Deenanath grass. In early summer, for better forage production, rice bean can be intercropped between the widely spaced perennial grasses such as, hybrid bajra Napier and Guinea grass, which make slow growth and do not show mutual competition. Rice bean can also be given with Anjan, Marvel, Sehima grass, etc.

14.22 CROP SEQUENCE

It is best suited as a short rotation rainy season forage crop for 300% fodder crop intensity. It may be followed by berseem or oats in the winter season and maize, *Pennisetum pedicellatum*, or *jowar* in the summer season. It may also be intercropped with hybrid *Pennisetum* during summer season after the harvest of intercropped berseem or oats. Growing of legumes of the *Vigna* group after rice bean aggravates diseases in the crop following rice bean.

14.23 HARVESTING

Rice bean crops are photoperiod sensitive and flower only in December under the influence of short days, when there is acute fodder shortage in a dairy farm. February to May sown crop is harvested in July and August when it attains the maximum vegetative growth. July and August sown crops are harvested at flowering or a bit late in December. For fodder, it can be cut 70–90 days after sowing between 50% flowering and pod initiation stage. Rice bean has poor regeneration capacity. Cutting at 25 cm above ground level is best and regrowth is maximum because more axillary buds and leaves are left on stubbles maintaining higher total carbohydrates. Only one cut is usually taken and the best time for harvest is early pod stage.

14.24 GREEN FODDER YIELDS

Green forage yield may be obtained 175–350 quintals/ha from summer crop and 200–350 quintals/ha from kharif crop. Under uninterrupted growth, the crop can yield 80 quintals/ha dry matter. Dry matter production of about 53–68 quintals/ha is obtained during May to June and 80–94 quintals/ha during November to December. Average grain yield of this crop ranges from 13–25 quintals/ha. The calcium and iron contents are low in rice bean. The seeds are free from any toxic cyanogenetic compounds. In India and Nepal, rice bean is sown in February and March for harvest during summer and in July and August for harvest in December (Khanal et al., 2009). It can be sown alone in small fields or along bunds of rice terraces. Rice bean benefits from being sown between rows of a tall cereal such as maize or sorghum that it can use for climbing. Rice bean is a hardy plant that is resistant to many pests and diseases, and it does not require fertilizer or special care during growth. In Nepal, farmers clip the tips of the plant to promote pod formation. Rice bean usually matures in 120–150 days after sowing but may need more time at higher altitudes. Seeds are harvested when 75% of the pods turn brown. Harvesting is best done in the morning or late afternoon to reduce the risk of heat-induced shattering. After the harvest, the vines and pods remain on the ground for 2–3 days after which the plants are threshed. The crop residues can then be used as fodder (Khanal et al., 2009).

In India, late maturing and photosensitive landraces of rice bean are cultivated as a fodder crop. They are sown during long-day periods in order to prevent the plant from flowering. Dual-purpose varieties may be cut when the pods are half-grown, but the hay should be handled as little as possible because the leaves drop easily. In Bengal (India), fodder yields were reported to range from 5–7 t DM/ha in May and June to 8–9 t DM/ha in November and December (Chatterjee et al., 1977). Lower values have been reported: 5–6 t DM/ha in Myanmar (Tin Maung Aye, 2001) and 2.9 t DM/ha in the subhumid Pothwar plateau of Pakistan (Qamar et al., 2014). In India, rice bean grown with Nigeria grass (*P. pedicellatum*) yielded 7.6 t DM/ha after the application of 20 kg N/ha (Chatterjee et al., 1977). In Pakistan, rice bean grown with sorghum (50:50 mix) yielded up to 12 t dry matter (DM)/ha (Ayub et al., 2004).

Like other legume forages, fresh rice bean forage is relatively rich in protein, though its concentration is extremely variable (17–23% DM). Rice bean hay and straw are slightly less nutritious (16% and 14% protein in the DM, respectively). Rice bean forage is also rich in minerals (10% of the DM in the fresh forage), and particularly in calcium (up to 2% in the fresh forage). Rice bean straw contains large amounts of mineral matter (more than 20% of DM), though it is highly variable.

14.25 TOXICITIES

Rice bean seeds contain antinutritional factors. Antitrypsin activity is notable, though comparable to that of cowpea (*V. unguiculata*) and black gram (*V. mungo*). Hemagglutinin activity was found to be lower than that of the two latter legumes (Malhotra et al., 1988). Relatively low levels of phenols and phytic phosphorus have been reported (Gupta et al., 1992). Ruminants rice bean fodder (fresh, hay, and straw) and rice bean seeds can be fed to ruminants. Rice bean contains phytic acid, polyphenol, tannins, trypsin inhibitors, other antinutrients, and flatus producing oligosaccharides. However, the content of all these is lower than in many comparable pulses. The effects are strongly reduced through common cooking practices such as soaking, germination, dry roasting and cooking. It has also been discovered that rice bean does not contain any toxic or allergic components.

14.26 PALATABILITY

Rice bean forage at the pre-flowering stage is palatable to sheep. In Nepal, farmers have emphasized the softness and palatability of rice bean fodder for livestock (Joshi et al., 2008). Hay fodder is also palatable to all kind of animals. Rice bean straw was reported to be relished by cattle.

14.27 GREEN MANURE AND COVER CROP

Rice bean is nitrogen-fixing legume that improves the N status of the soil, thus providing N to the following crop and enhances the fertility of soil to a great extent. Its taproot has a beneficial effect on soil structure and, when plowed in, returns organic matter and N to the soil. Rice bean grown before or after a rice or maize crop is beneficial. In Thailand, it is profitably sown between the rows of maize once the crop has reached maturity, but before harvest so that rice bean covers enough soil at harvest. It is then possible to harvest the rice bean, thresh to obtain the seeds, and bring the dry plants back to the field where they provide soil cover for the dry season. When intercropped, it increases the yield of maize and other staple crops. It also provides a stealthily good soil cover and reduces erosion, and is often planted on terrace risers. The plant residue is valuable livestock fodder and is well known to increase milk production. Rice bean is commonly grown on marginal and exhausted soil, fixing nitrogen, and replenishing the nutrient balance. In Eastern Nepal, it is cultivated in rotation with ginger and other cash crops, for the benefit of soil nutrients and to reduce pests and diseases.

14.28 UTILIZATIONS

Rice bean is a multipurpose legume, sometimes considered as neglected and underutilized (Joshi et al., 2008). All parts of the rice bean plant are edible and used in culinary preparations. The dry seeds can be boiled and eaten with rice or they can replace rice in stews or soups. In Madagascar, they are ground to make nutritive flour included in the food for children. Unlike other pulses, rice beans are not easily processed into dhal, due to their fibrous mucilage that prevents hulling and separation of the cotyledons. Young pods, leaves, and sprouted seeds are boiled and eaten as vegetables. Young pods are sometimes eaten raw. Rice bean is useful for livestock feeding. The vegetative parts can be fed fresh or made into hay and the seeds are used as fodder. Before feeding, the woody portions and soiled or mildewed parts of the straw should be removed. In the marginal hills of Nepal, farmers consider rice bean both as a grain and fodder legume and look for dual-purpose landraces (Khanal et al., 2009). Rice bean is a practical soil-improving, nitrogen-fixing, green manure crop. Stems can be cut and used as mulch and stock feed. Hens enjoy the green leaves. Rice bean is a valuable legume for erosion control on steep slopes and banks. Use this hardy, easy to grow legume as a weed smothered. In Asia, the bean is planted in rice fields after rice harvests, to add nitrogen and humus to benefit. Rice bean is nutritionally rich legume, but despite its nutritional excellence, it has been put in underutilized category. Because of this and several other reasons, the people are not aware of its nutritional benefits. Moreover, the complete nutritional details are also not available on this pulse. Rice bean is grown for green manure, as a cover crop, and used as a living fence or biological barrier. Rice bean is fed to livestock as green forage or hay.

14.29 NUTRITIVE VALUE

Rice bean is a palatable and highly nutritious fodder. It is richer than cowpea and black gram in protein, calcium, and phosphorus. The chemical composition of the fodder (in percentage) is as follows (Table 14.1).

TABLE 14.1 Chemical Composition of Rice Bean.

CP	EE	CF	TA	NFE
16.9	1.9	30.6	7.8	43.8

The digestibility coefficients for dry matter, crude protein, ether extract, crude fiber, and nitrogen free extracts are 55.05%, 63.08%, 67.45%, 36.82%, and 72.58%, respectively. The digestible crude protein and total digestible nitrogen values are 8.06% and 52.84%, respectively. Nitrogen and calcium are positively balanced at flowering, whereas phosphorus is slightly negatively balanced requiring supplementation. The seed contains 14–24% of protein. It contains a high amount of vitamins, thiamine, niacin, and riboflavin.

FIGURE 14.1 **(See color insert.)** Rice bean.

KEYWORDS

- **rice bean**
- **agronomic management**
- **fodder yield**
- **nutritive value**
- **utilization**

REFERENCES

Ayub, M.; Tanveer, A.; Nadeem, M. A.; Shah, S. M. A. Studies on the Fodder Yield and Quality of Sorghum Grown Alone and in Mixture with Rice Bean. *Pak. J. Life Soc. Sci.* **2004,** *2* (1), 46–48.

Chatterjee, B. N.; Dana, S. Rice Bean. *Tropical Grain Legume Bulletin No.1*; International Grain Legislative Information Center, 110 Legistative Building:Olympia, 1977.

Ecoport, 2014. Ecoport database. Ecoport 2014. https://www.feedipedia.org/node/234, (accessed Feb 12,2017).

Gupta, J. J.; Yadav, B. P. S.; Gupta, H. K. Rice Bean (*Vigna umbellata*) as Poultry Feed. *Indian J. Anim. Nutr*. **1992,** *9* (1), 59–62.

Joshi, K. D.; Bhanduri, B.; Gautam, R.; Bajracharya, J.; Hollington, P. B. Rice Bean: A Multi-purpose Underutilized Legume. In *New Crops and Uses: Their Role in a Rapidly Changing World*; Smartt, J., Haq, N., Eds.; CUC: UK, 2008, pp 234–248.

Khadka, K.; Acharya, B. D. *Cultivation Practices of Rice Bean,* 1st ed.; Local Initiatives for Biodiversity, Research and Development (LI-BIRD): Pokhara, Nepal, 2009.

Khanal, A. R.; Khadka, K.; Poudel, I.; Joshi, K. D.; Hollington, P. *The Rice Bean Network: Farmers Indigenous Knowledge of Rice Bean in Nepal* (*report N°4*)*, EC.* Report on Farmers' Local Knowledge Associated with the Production, Utilization and Diversity of Rice Bean (*Vigna umbellata*) in Nepal. 6th FP, Project no. 032055, FOSRIN (Food Security Through Rice Bean Research in India and Nepal). 2009.

Malhotra, S.; Malik, D.; Dhindsa, K. S. Proximate Composition and Antinutritional Factors in Rice Bean (*Vigna umbellata*). *Plant Foods Hum. Nutr.* **1988,** *38* (1), 75–81.

Qamar, I. A.; Maqsood Ahmad; Gulshan Riaz; Khan, S. Performance of Summer Forage Legumes and Their Residual Effect on Subsequent Oat Crop in Subtropical Sub Humid Pothwar, Pakistan. *Pakistan J. Agric. Res*. **2014,** *27* (1), 14–20.

Rajerison, R.; *Vigna umbellata* (Thunb.) Ohwi & H. Ohashi. In *1: Cereals and pulses/ Céréales et légumes secs*; *Brink, M., Belay, G. Eds.*; PROTA; Wageningen, Pays Bas. **2006**.

Tateishi, Y. A revision of the Azuki bean group, the subgenus *Ceratotropis* of the genus *Vigna* (Leguminosae). Ph.D. Thesis, Tohoku University, Sendai, Japan. 1985.

Tin Maung Aye. Developing Sustainable Soil Fertility in Southern Shan State of Myanmar. PhD Thesis. Massey University, Palmerston North, New Zealand. 2001.
Verdcourt, B. Studies in the Leguminosae: Papilionoideae for the "Flora of Tropical East Africa": IV. *Kew Bull.* **1970,** *24*, 558–560.

CHAPTER 15

BERSEEM (EGYPTIAN CLOVER)

ARUN KUMAR BARIK[1*] and MD. HEDAYETULLAH[2]

[1]*Department of Agronomy, Palli Siksha Bhavana, Institute of Agriculture, Visva Bharati University, Sriniketan 731236, West Bengal, India*

[2]*Department of Agronomy, Faculty of Agriculture, Bidhan Chandra Krishi Viswavidyalaya, Mohanpur 741252, West Bengal, India*

[*]*Corresponding author. E-mail: akbarikpsbvb@rediffmail.com*

ABSTRACT

Berseem or Egyptian clover is winter forage popularly known as the king of fodder crops. It is available for 6–7 months from November to May and it gives four to six cuts during winter, spring, and early summer seasons and provides nutrition, succulent, and palatable forage. Cultivation of berseem has got a soil building characteristics and improves the physical, chemical, and biological properties of the soil resulting in better growth and yield of crops in rotation. Thus, berseem is very important forage crop from the view point of conservation farming and imparts sustainability to soil productivity and crop production system. Forage yield potential of berseem crop is very high because of its good regrowth capacity. The green fodder yield is about 1000–1200 quintals/ha under improved agronomic management practices and favorable weather conditions. Five to six cuttings can be done under irrigation and one or two at the end of the cool season in dry land situation. The green forage of berseem on dry matter basis contains 12.8% crude protein, 26.7% crude fiber, 10.6% ash, 35.8% nitrogen free extract, and 1.4% ether extract.

15.1 SCIENTIFIC CLASSIFICATION

Kingdom: Plantae
Orders: Fabales
Family: Fabaceae
Genus: *Trifolium*
Species: *alexandrinum*
Binomial name: *Trifolium alexandrinum* (L.)

15.2 BOTANICAL NAME

Trifolium alexandrinum (L.)

15.3 COMMON NAMES

Berseem, berseem clover, Egyptian clover

15.4 INTRODUCTION

Berseem or Egyptian clover (*Trifolium alexandrinum L.)* is winter forage popularly known as the king of fodder crops. It is available for 6–7 months from November to May and it gives four to six cuts during winter, spring, and early summer seasons and provides nutrition, succulent, and palatable forage. The green forage can also be converted into excellent hay and utilized for enrichment of poor-quality roughages such as straw. Cultivation of berseem has got a soil building characteristics and improves the physical, chemical, and biological properties of the soil resulting in better growth and yield of crops in rotation. Thus, berseem is very important forage crop from the view point of conservation farming and imparts sustainability to soil productivity and crop production system. Berseem is one of the most important leguminous forages in the Mediterranean region and in the Middle East. Berseem is a fast-growing, high-quality forage that is mainly cut and fed as green chopped forage. It is often compared to alfalfa, due to its comparable feed value. However, unlike alfalfa, it has never been reported to cause bloat. It is slightly less drought resistant but does better on high moisture and alkaline soils. Moreover, berseem can be sown in early autumn and can thus provide feed before and during the colder months (Suttie, 1999). It is very productive when temperatures rise after winter (Hannaway et al., 2004). The seeds are abundant under favorable conditions. Berseem can be made into silage with oats or be fed chaffed and mixed with chopped straw (Hannaway et al., 2004).

15.5 ORIGIN AND HISTORY

Berseem probably originated in Syria. It was introduced into Egypt in the sixth century (Hannaway et al., 2004). It was introduced into India from Egypt in the 1904 and into Pakistan, South Africa, the United States, and Australia in the 20th century. Berseem has been one of the fastest spreading fodder species in recent times, mainly under small-scale farming conditions (Suttie, 1999).

15.6 AREA AND DISTRIBUTION

Berseem is grown only in few countries of the world. It is basically cultivated in Egypt, Syria, Pakistan, and India. Berseem green forage is used for horses, cattle, camels, donkeys, etc. It is grown in irrigated areas of India such as Punjab, Haryana, Delhi, Rajasthan, Uttar Pradesh, and some parts of Bihar, Maharashtra, and Andhra Pradesh. Cultivated areas reached 1.3 million hectare in 2007 in Egypt (El-Nahrawy, 2011) and 1.9 million hectare in India (ICAR, 2012). Morocco adopted berseem in the beginning of the 20th century and 50,000 ha were grown under irrigation in 2005 (Merabet et al., 2005).

15.7 PLANT CHARACTERISTICS

Berseem is an annual, sparsely hairy, erect forage legume, 30–80-cm high (Hackney et al., 2007). Berseem plant root is a shallow taproot system. Its stems are succulent, hollow, branching at the base. Leaves are small, trifoliate tender, and slightly hairy on upper surface. Flowers are yellowish-white and form dense, elliptical clustered heads about 2 cm in diameter. The flowers must be cross-pollinated by honeybees to produce seeds. The fruit is a pod containing one single white to purplish-red seed, small in size (2 mm), and egg shaped (Smoloak et al., 2006).

15.8 CLIMATIC REQUIREMENTS

Berseem requires a dry and cool and moderately cold climate. Such conditions prevail during winter and spring seasons in North India which is considered as favorable and productive zone for this crop. The optimum temperature at the time of sowing berseem is 25°C. For vegetative growth, temperature range of 25–27°C has been found ideal. Uniformly high temperature in South Indian conditions limits the cultivation of berseem. Berseem is mainly valued as a winter crop in the subtropics as it grows well in mild winter and recovers strongly after cutting. It does not grow well under hot summer conditions. It is cultivated from 35°N to the tropics, from sea level up to 750 m (Hannaway et al., 2004). Berseem has some frost tolerance up to −15°C for some cultivars (Suttie, 1999). During frost plant remains dormant and no regeneration is recorded. Berseem can grow in areas where annual rainfall ranges between 550 and 750 mm. It can withstand some drought and short periods of waterlogging.

15.9 SOIL AND ITS PREPARATION

Well-drained clay to clay loam soils rich in humus, calcium, and phosphorus are suitable for good crop of berseem. However, it can be grown on sandy loam soil but requires frequent irrigations. Comparatively, heavy-textured soils considered better due to greater water retaining capacity.

The land should be well leveled to obtain even distribution of irrigation water and to avoid water stagnation, berseem can be grown in saline sodic soils if salt concentration is not allowed to accumulate above certain critical level through field flooding and leaching to provide optimum conditions for seed germination and crop establishment. Once established, the crop can tolerate fair amount of salt concentration. It is moderately tolerant of salinity and can grow on a wide range of soils, though it prefers fertile, loamy to clay soils with mildly acidic to slightly alkaline pH (6.5–8) (Hackney et al., 2007).

15.10 SEEDBED PREPARATION

The land should be plowed with mould board plow followed by two to three operations by desi plow/cultivator. The preparation of good seedbed is an essential component of cultivation practices to obtain desired level of tilth. Fine seedbed is required especially when berseem is to be grown as seed crop in rows without puddling to facilitate weed removal and rouging for quality seed production. When the crop is to be sown in puddle beds, thorough cultivation is not required, only cross-harrowing/plowing is needed to remove established weeds, stubbles of the previous crop.

15.11 VARIETIES

Berseem is broadly classified into two groups, namely, diploid and tetraploid. In diploid group, most promising varieties are Mescavi, Berseem Ludhiana 1, Chhindwara, and IGFRI 99-1. These varieties are very much susceptible to low and high temperatures. In tetraploid group most promising varieties are Pusa Giant and T 678.

15.11.1 DIPLOID VARIETIES

15.11.1.1 MESCAVI

It is a fast-growing variety and attains plant height of about 75 cm at flower initiation stage. On an average, it gives 500–600 quintals green fodder and 100–125 quintals dry matter yields per hectare in about five cuttings. It contains about 20% crude protein (CP) on dry matter basis at early flowering stage.

15.11.1.2 BL-1

This is a long-duration variety as compared to Mescavi. Because of this, one additional cutting may be obtained from this variety by the end of June. It gives, on an average, green fodder and dry matter yields of 600 and 130 quintals/ha, respectively.

15.11.1.3 BL-22

This is a long-duration variety which gives additional cut during June. It gives, on an average, green fodder and dry matter yields of 750 and 135 quintals/ha, respectively.

15.11.2 TETRAPLOID VARIETIES

15.11.2.1 PUSA GIANT

It is an autotetraploid variety developed at IARI, New Delhi. It has dark green broader and thicker leaves. Seeds are bold (4.12 g/1000 seeds). Its yield potentiality is 10–15% higher than Mescavi variety. It is frost and winter hardiness resistant.

Other improved varieties of berseem are T 724, T 560, JHB 46, UJB 3, JB 4, JB 1, JB 2, S 99-1, and VPB 1.

15.12 SOWING TIME

The best time for sowing berseem in North India is the first fortnight of October. Early sown crop is may be adversely affected due to heavy rains and weed infestation. Late sown berseem also suffers due to frost and low temperature. Sowing time is an important factor governing germination, seedling survival, number of cuts, and herbage production. Berseem should be sown when the temperature is in the range of 25–27°C. Thus, the optimum sowing time of berseem in Punjab, Haryana, and Uttar Pradesh is the entire month of October. In Bengal and Gujarat, sowing is taken up only in the month of November. Sowing can continue up to first week of December in eastern region. Delay in sowing results in loss or one of two cuttings. Timely sowing extends the period of forage availability and thereby increases the total yield.

15.13 SEED TREATMENT AND INOCULATION

To treat the seed with fungicide, apply Thiram @ 3 g/kg of seed. Seeds should be inoculated with *Rhizobium trifolii* bacteria culture which helps in nitrogen fixation before sowing. If bacteria culture is not available, then collect 500 kg soil from where berseem previously was grown and broadcast uniformly to the new field. Apply 20 g *rhizobium* for 1 kg of seeds before sowing for better nodulation. Berseem is a leguminous crop which enriches the soil with sizeable quantities of nitrogen through symbiotic nitrogen fixation with the help of Rhizobium bacteria. Therefore, berseem seed should be inoculated with culture of *R. trifolii* to enhance the process of biological nitrogen fixation in root nodules, especially in soils where berseem is being grown for the first time.

15.14 SEED RATE

The optimum seed rate of timely sown berseem is 25–30 kg/ha. For late and early sown seed required is 35 kg/ha. For better growth and green forage yield, diploid and tetraploid varieties are mixed together in 1:1 or 1:2 ratio. Under normal condition, the optimum seed rate of berseem has been found to be 25 kg/ha. When the sowing is taken up earlier than the appropriate time, the quantity of seed used is increased by 15–20% to compensate the loss of seedling mortality occurring due to prevailing high atmospheric temperature.

15.15 SOWING METHOD

There are two berseem sowing methods, namely, dry and wet bed methods.

15.15.1 DRY METHOD

In dry bed method, seeds are directly broadcasted in well-prepared field and covered 1 cm with soil with the help of wooden plank. In dry sowing method, irrigation should be given only after proper germination.

15.15.2 WET METHOD

In this method, field is plowed thoroughly in standing water to make impervious layer. The seedbed for berseem sowing is prepared by filling the water to a depth of 4–5 cm, raking the soil and creating the muddy condition by light puddling. Puddling operation is done by puddler or by wooden plank. After complete puddling, seeds are sown immediately in soil water suspension. Overnight soaked seeds are broadcast in standing muddy water in crosswise directions to obtain uniform seed distribution. For better seed establishment, soil water suspension or mud is very important practice in berseem. The seeds are generally broadcast manually by hand uniformly. For getting good quality fodder, mix 2 kg of mustard seed along with full rate of berseem seed. The sowing should be done toward the evening or during nonwindy periods of the day.

15.16 SEED CLEANING

Chicory (*Chicorium intybus*) is very problematic weed in berseem field. Chicory is found admixed with berseem seed. Since the size of chicory seed resembles with berseem seed, it becomes difficult to separate them by ordinary methods. However, the seed of berseem is oval while the seed of it is conical. To remove chicory seeds, 10% common salt solution is used. The chicory seeds being lighter in weight than berseem seed float on the surface while berseem seeds settle down at the bottom of container. In this way, chicory seeds may be drained off and berseem seed collected.

15.17 SPACING

For line sowing, row to row spacing is maintained about 25–30 cm, but depending upon the soil, climate, and irrigation facilities it may be increased. In dry areas, the spacing may be increased up to 50 cm when it is grown in intercropping systems with oats and barley.

15.18 CROPPING SYSTEMS

Berseem can be fit in sequential cropping systems with annual grain crops, in overlapping cropping systems with perennial grasses, and in parallel cropping system with rabi grain crops. Berseem grown in sequence and overlapping cropping systems imply to intensive forage production systems in milkshed areas to meet the forage requirement round the year. These systems are adopted on specialized dairy farms to harvest green nutritious forage with the objectives of stabilizing milk production over the periods, reducing the concentrate mixture on animal ration, and economizing on use of fertilizer nitrogen in cropping systems. The parallel cropping system may be adopted by small farmers who intend to produce food and forage from the limited land area.

The advantages of berseem in intensive forage cropping system are growing. Berseem ensures effective utilization of land during dormant phase of perennial grass component. Combination cropping of berseem and grasses helps in balanced utilization of plant nutrients from different soil depths. Berseem improves physical, chemical, and biological properties of soil and thus improves its fertility status. Berseem in winter acts as live mulch and protects the grass tussocks from damage by prom age by providing moist soil conditions. The system provides opportunities for rational use of water as extra irrigation is not required for the establishment of grass in standing berseem crop. Intercropping perennial grasses with forage legumes has been reported to reduce antiquality constituents such as oxalates in hybrid Napier, besides providing balanced and nutritious herbage to animals. There is almost continuous flow of green forage throughout the year from the same piece of land which is important for farmers with small size of holding.

Some of the important rotation of the berseem is given below:

- Paddy–berseem–cowpea
- Maize–berseem–maize
- Sorghum–berseem–maize
- Pearl millet–berseem–maize + cowpea
- Napier–berseem–cowpea
- Teosinte–berseem–maize+ cowpea
- Napier grass + berseem

15.19 NUTRIENT MANAGEMENT (MANURES AND FERTILIZERS)

Organic manures always give better environment to the soil and plant too for long term. For obtaining good plant health and high fodder production, add about 10–15 t/ha farm yard manure or compost. Berseem being a legume crop which fixes the atmospheric nitrogen in soil through symbiotic *R. trifolii* bacteria. However, a light dose of nitrogen is required for initial plant establishment. Therefore, fertilizer nitrogen is required only for establishment prior to the formation of root nodules. In irrigated farming, fertilizer use is the most important factor-influencing growth and productivity of the crops. Experimental evidences have proved that 20 kg N/ha at sowing is the optimum dose. When the crop is raised on poor soils without inoculums, top dressing of 10 kg N/ha is done after each cut in addition to 30 kg N/ha basal dose to encourage good regeneration, quick growth, and high yield. In general, the responses to applied phosphorus vary widely with available soil phosphorus, soil pH, phosphate sources, application method, water supply, and crop duration. The placement of phosphorus in the rhizosphere is desirable to reduce its fixation and to increase availability to young seedling. Placement of phosphorus has been found superior to broadcast in terms of forage yield and phosphorus uptake. The crop responds significantly up to 80–90 kg P_2O_5/ha. Phosphorus is essential at initial crop stage, basal application proved better than top dressing. The potassium requirement of berseem has been found to be 30–40 kg K_2O/ha in low-potassium soils.

15.20 WATER MANAGEMENT

Berseem requires huge quantities of water for producing high succulent biomass. For every kilogram of plant dry matter produced as much as 500 kg of more or water may be necessary in a dry climate. Therefore, adequate and timely water supply is one the basic inputs for obtaining higher green forage yield. Irrigated berseem must be sown earlier and irrigated on a weekly basis at the beginning. Ten to 15 irrigations are generally necessary for fodder production (Suttie, 1999). First irrigation may be applied within 3–5 days on light soil, whereas in clay or heavy soils, it may be applied within 8–10 days. Subsequently, irrigate the crop every 15–20-day interval depending upon weather and soil moisture conditions.

15.21 INTERCULTURE OPERATION AND WEED MANAGEMENT

The weed management is one of the important operations of berseem production. The major associated weed of berseem crop is chicory

(*C. intybus*). The nature of this weed is such that it infests from field to seed and vice versa. The intensity of field infestations could be minimized by treatment with 10% solution of common salt and deep summer plowing with soil inversion plow after final harvest of the crop. Berseem sown in mixture with oats or ryegrass smothers weeds during establishment, and regrows after cutting at the time when the oats are harvested (Clark, 2008).

15.22 INSECT-PEST AND DISEASE MANAGEMENT

During the month of December and January when the crop gives good vegetative growth and cloudy days persist for longer period, the heavy infestation of fungal diseases such as root rot caused by *Rhizoctenia soloni* and *Fusanlum smitactum* and stem rot caused by *Sclerotinia trifoliorum* occur. This disease has been observed more acute under field is heavily manured with undecomposed farmyard manure, water stagnated creation damp conditions, light penetration at the ground is curtailed due to delayed cutting, and cloudy condition prevails for longer period.

Agronomic approaches to minimize the predominance such as avoiding the growing of berseem crop in the same field year after year and deep plowing during summer. Use well-rotten manure in proper quantities. Fertilize the crop with heavy dose of potassium. Level the field properly to avoid water stagnation. Avoid too frequent irrigations during cloudy days. Cut the crop frequently to expose the ground for adequate light availability.

15.23 HARVESTING

The first cutting should be taken at 60 days after sowing. Subsequent cuttings can take at 25–35-day interval depending upon the regrowth. Height of cutting should be within 5–7 cm from the surface for quick regrowth. Berseem is highly nutritious, succulent, and palatable forage for all types of livestock. It stimulates milk production of cows and buffaloes and is popular both for milch and draught animals. Berseem is a good source of CP, calcium, phosphorus, and ether extract (EE).

15.24 PASTURE

Berseem does not perform well under grazing as livestock may damage its upper growing buds. When grazing is intended in irrigated pastures, it should start before the sward becomes too erect. The sward should be grazed down to 5–6-cm high and the rest period should be about 30–40 days between grazing periods (Hackney et al., 2007).

15.25 GREEN FORAGE

It is high-quality green forage. Berseem should be cut 50–60 days after planting and then every 30–40 days (Suttie, 1999). The highest yield of protein with a relatively low yield of fiber was obtained by cutting the plant at a height of about 40 cm. Five to six cuttings can be done under irrigation and one or two at the end of the cool season in dry land situation.

15.26 HAY

Berseem is not well suited to make hay because its succulent stems do not dry easily. When berseem is intended for hay, only the last spring cut should be used as it is drier. It may also be useful to wilt berseem in the field and then let it dry on roof tops to make it into hay (Suttie, 1999).

15.27 SILAGE

Berseem fodder makes good quality silage. Berseem can be mixed with 20% ground maize to provide high quality silage. It is possible to make silage with berseem and 5% molasses.

15.28 CUTTING MANAGEMENT AND FORAGE YIELD

Forge yield potential of berseem crop is very high because of its good regrowth capacity. The green fodder yield is about 1000–1200 quintals/ha under improved agronomic management practices and favorable weather conditions. Mixing Japan rape or Chinese cabbage 2.25-kg seed/

ha increases the yield by 20–25% in first cut. The yield may further be increased by introducing early cutting.

15.29 SEED PRODUCTION

Take the final cut in first week of March in dry areas and in the third week of March in humid areas. All weeds should be removed from the field. Irrigate the crop when it is required for good plant stand. The seeds mature in the month of May–June. The average seed yield of berseem is 40–50 quintals/ha.

15.30 NUTRITIVE VALUE

Berseem is a highly nutritious fodder for all types of livestock. The green forage of berseem on dry matter basis contains 12.8% CP, 26.7% crude fiber, 10.6% ash, 35.8% nitrogen-free extract, and 1.4% EE. In green fodder, digestibility of CP is 81%, EE 50%, and nitrogen free extract 80% (Chatterjee and Das, 1989).

15.31 UTILIZATION

Grazing is possible, though less common than cutting. Berseem is widely used green fodder worldwide for its nutritive value and its palatability. Berseem can be converted into good hay. Powdered berseem hay can also mixed with concentrates for poultry birds. It is possible to make silage with berseem and 5% molasses. Berseem clover can also be used as green manure crop (Hannaway et al., 2004).

15.32 TOXICITIES

Berseem fodder contains estrogenous substances. Excessive feeding of berseem in the young stage or grazing on it when it is wet with dew in the morning causes bloat (tympanitis). If animals are allowed to graze or feed on berseem, it should be sprayed with linseed or mustard oil emulsion to prevent bloat.

FIGURE 15.1 (See color insert.) Berseem.

KEYWORDS

- **berseem**
- **agronomic management**
- **fodder yield**
- **nutritive value**
- **toxicities**
- **utilization**

REFERENCES

Chatterjee, B. N.; Das, P. K. *Forage Crop Production Principles and Practices*; Oxford & IBH Publishing Co. Pvt. Ltd.: Kolkata, 1989.

Clark, A., Ed. Berseem. In *Managing Cover Crops Profitably*; Diane Publishing, 2008.
El-Nahrawy, M. A. *Country Pasture/Forage Resource Profiles;* FAO Rome: Egypt, 2011.
Hackney, B.; Dear, B.; Crocker, G. *Berseem Clover*; New South Wales Department of Primary Industries, Primefacts, No.388, 2007.
Hannaway, D. B.; Larson, C. *Berseem Clover* (*Trifolium alexandrinum* L.); Species Selection Information System: Oregon State University, 2004.
ICAR. Forage Crops and Grasses. In *Handbook of Agriculture*, 6th ed.; ICAR, 2012.
Merabet, B. A.; Abdelguerfi, A.; Bassaid, F.; Daoud Y. Production and Forage Quality of Berseem Clover According to the Water Supply in Mitidja (Algeria). *Fourrages* **2005,** *181*, 179–191.
Suttie, J. M. Berseem Clover (*Trifolium alexandrinum*), In *A Searchable Catalogue of Grass and Forage Legumes*; FAO, 1999.

CHAPTER 16

LUCERNE (ALFALFA)

PARTHA SARATHI PATRA* and TARUN PAUL

Department of Agronomy, Uttar Banga Krishi Viswavidyalaya, Pundibari 736165, Cooch Behar, West Bengal, India

Corresponding author. E-mail: parthaagro@gmail.com

ABSTRACT

Lucerne is a perennial plant and may supply green fodder continuously for 3–4 years from the same sowing. It grows to a wide range of climatic conditions ranging from tropical to alpine. Lucerne is grown in winter season and has been grown successfully in arable cropping regions. Being a deep-rooted crop, it extracts water from the deeper zone of the soil. It can be raised both as rainfed or irrigated crop in high water table areas. It is a deep-rooting crop which can sustain dry matter production at times of low rainfall. The average green fodder yield of lucerne varies from 800–1000 quintals/ha. The risk of producing bloat in animals is very real when cattle are grazed. Bloat causes due to presence of certain glucosides known as saponins in lucerne plant at green stage.

16.1 BOTANICAL CLASSIFICATION

Kingdom: Plantae
Order: Fabales
Family: Fabaceae
Sub-family: Faboideae
Genus: *Medicago*
Species: *sativa* L.
Binomial name: *Medicago sativa* L.

16.2 COMMON NAMES

Alfalfa and lucerne.

16.3 BOTANICAL NAME

Medicago sativa L.

16.4 INTRODUCTION

Lucerne or alfalfa is a forage legume that has been cultivated for around 2000 years. It is a perennial plant and may supply green fodder continuously for 3–4 years from the same sowing. It grows to a wide range of climatic conditions ranging from tropical to alpine. Lucerne is grown in winter season and has been grown successfully in arable cropping regions. Being a deep-rooted crop, it extracts water from the deeper zone of the soil. It can be raised both as rainfed or irrigated crop in high water table areas. It is a deep-rooting crop which can sustain dry matter (DM) production at times of low rainfall. We must expect rising temperatures due to climate change and drought conditions are likely in the future.

Lucerne is relished by all kinds of livestock, because it yields nutritious and palatable green fodder, which possesses about 16–25% crude protein (with 72% digestibility) and 20–30% fiber. It is naturally high in many essential vitamins and minerals, including A, D, E, K, and B vitamins; biotin, calcium (1.5%), folic acid, iron, magnesium, potassium, and many others, especially when dried. Owing to its high protein, vitamin, and mineral content, it is also included as a feed component in poultry and piggery. It can also be easily converted into silage and hay. Lucerne supplies green fodder for a longer period (November–June) in comparison to berseem (December–April).

16.5 MAIN REASONS FOR GROWING LUCERNE

1. It produces green fodder yield on regular basis and during the dry summer it performs better than other forages.

2. It gives a very high DM yield for longer times (3–4 years).
3. It is highly proteinaceous forage crop, protein content ranges from 15% to 20% based on growth stages.
4. It is an excellent cutting fodder crop and is complementary to other forage crops.
5. It is rich in vitamins and minerals, and if cut at the right stage is low in fiber and high in energy.
6. It fixes atmospheric nitrogen and thereby improves soil fertility status.

16.6 ECONOMIC IMPORTANCE

1. Lucerne supplies green fodder for livestock feeding.
2. Lucerne is made into hay and silage for milk production. For hay making it is excellent fodder.
3. On drying, it is made into high-protein meal or pellets. Dried lucerne pellets have high carotene content, high energy and protein (up to 20%), and are a valuable feed for livestock.
4. Further benefits from growing lucerne derive from its remarkable deep taproot system. The roots penetrate to a depth of 3 m, and thus enabling the plant to draw moisture and minerals from a considerable depth.
5. The decomposition of plant biomass contributes to the fertility by increasing the humus content.
6. Nitrate leaching is also reduced as the plants take up a large amount of water during growth, thus reducing runoff.

16.7 ORIGIN AND HISTORY

Lucerne is one of the oldest cultivated fodder crops in the world. It is generally believed that lucerne originated in Southwest Asia. It was first cultivated in Persia (Iran), the name alfalfa being an Arabic word. From Iran it was taken to Greece in about 500 B.C. and from there it spread to Italy. The Spaniards introduced it to America. Lucerne was introduced in India from northwest sometime in 1900. It has now become very popular forage crop in the world.

16.8 GEOGRAPHIC DISTRIBUTION

Lucerne is grown all throughout the world. The major lucerne producing countries are United States, Canada, Argentina, India, Australia, New Zealand, France, Italy, and Russia.

16.9 BOTANICAL DESCRIPTION

The plant parts of lucerne are described below.

16.9.1 ROOT SYSTEM

Lucerne has a deep root system consisting of strong main taproot and number of lateral roots that makes the plant to draw moisture from the deeper soil layer and makes it drought tolerant.

16.9.2 STEM

The stem is erect and the branches arise from the crown, which is a woody base on stem near ground level. The number of branches may be as high as 40.

16.9.3 LEAVES

The leaves are trifoliate; the middle leaflet possesses a short petiole, a characteristic which distinguishes it from berseem. The long leaflets are sharply toothed on upper one-third of margin.

16.9.4 FLOWERS

The flowers are usually purple, but it may be blue, yellow, or white. They are fertilized by insects, especially bees.

16.9.5 SEEDS

The seeds are kidney shaped, very light in weight, and yellowish brown with a shiny surface.

16.10 ECOLOGICAL REQUIREMENTS

Lucerne is a native of temperate regions of Southwest Asia, but it is raised successfully even in most of the countries of the tropics. It performs well in cooler and dry climate than cloudy, humid, and wet conditions. It can also be raised in some regions below sea level, as well as at elevations of 2500 m altitude. High temperatures accompanied with high humidity drastically reduce DM yield. It can tolerate heat as well as fairly low temperatures (Singh et al., 2011).

16.11 SOIL

Lucerne needs sandy loam to clayey soils which are rich in organic matter, calcium, phosphorus, and potash. In heavy soils, lucerne grows with efficient drainage system as the crop does not tolerate waterlogging. It cannot thrive on alkaline soils but can be grown on acid soils with liberal application of lime.

16.12 FIELD PREPARATION

Lucerne needs well-leveled, very fine seedbed, as the seeds are very small in size. Therefore, field should be prepared thoroughly. One deep plowing with two to three harrowings followed by planking is sufficient. A fine seedbed ensures better contact of seeds with soil particles and facilitates quick and better germination.

16.13 SEED INOCULATION

Lucerne seed is inoculated with *Rhizobium meliloti* culture @ 200 g/10 kg seeds. *R. meliloti* improves nitrogen fixation in root nodules and thereby increases DM yield of lucerne. Soil nitrogen status is also improved.

16.14 SEED RATE AND METHOD OF SOWING

In case of broadcast method, a seed rate of 20–25 kg/ha should be used. In this method, care should be taken to cover the seed with 1–2-cm layer of soil. Seed may be broadcasted after last harrowing and then covered with planking. Care should be taken that seed should not go more than 1-cm deep as seed size of lucerne is very small. Line sowing needs only 12–15-kg/ha seeds but in case of intercropping, it requires only 6–12 kg/ha.

In the first cut, to obtain high green forage yield, lucerne seed should be mixed with mustard seed @ 1.5 kg/ha. Oat seeds @ 35 kg/ha are also broadcasted and mix it in soil with a cultivator before sowing lucerne.

16.15 TIME OF SOWING

Lucerne can be grown from the end of September to early December, but the best time for sowing lucerne is middle of October.

16.16 NUTRIENT MANAGEMENT

Being a legume crop, it fixes the atmospheric nitrogen in soil through symbiotic bacteria. Seed inoculation with *R. meliloti* is promising for crop performance, especially in soils where lucerne is being cultivated for the first time. Besides as starter dose of 20 kg N/ha, 60–75 kg P_2O_5/ha and 40 kg K_2O/ha are also applied at the time of sowing. In soils with low organic matter content, it is beneficial to apply farm yard manure @ 20 t/ha every year to maintain fertility status of soil (Singh et al., 2011).

16.17 WATER MANAGEMENT

Before sowing, one pre-sowing irrigation (*palewa*) is essential to obtain good germination. Lucerne requires frequent irrigations (7–10 days interval) at early stage of growth as it takes a long time to establish. Later on, this interval may be extended to 25–30 days as its root system gets well established. The crop requires about 15–20 irrigations in a year. Water requirement is quite high, being 858 L of water/kg of DM produced.

16.18 WEED MANAGEMENT

Lucerne takes longer time at early stage of growth to establish itself that facilitates more weed infestation up to the first cutting. For controlling weed infestation, first weeding should be done 20–25 days after sowing. In seed crop dodder *Cuscuta* is most important weed. It may reduce seed yield by 60%. For certified seed production of lucerne, its population should be <0.05% (20 *Cuscuta* seeds/kg lucerne seed). Preemergence application of pendimethalin @ 1–2 kg /ha or early postemergence (5–10 days after sowing) of Diquat @ 6–10 kg/ha effectively controls *Cuscuta*. "T 9" cultivar is found highly susceptible to this weed, while "LLC 6" and "LLC 7" are moderately tolerant to *Cuscuta* infestation. Presowing application of diuron @ 2.0 kg/ha or fluchloralin @ 1.0 kg/ha or EPTC @ 3.0 kg/ha or MCPB @ 0.75 kg/ha after 30 days after sowing or pronamide @ 1.0 kg/ha just after sowing controls the weeds in lucerne crop.

16.19 VARIETIES

Some of the improved varieties of lucerne commonly grown in India are given below.

16.19.1 SIRSA 8

This annual variety was developed at Fodder Research Station, Sirsa (Haryana). Its yield potential is about 35–40 t/ha of green fodder and 0.2–0.3 t/ha seed. It is suitable for Punjab, Haryana, Delhi, and Uttar Pradesh.

16.19.2 ANAND-3

Annual type, suitable for Himachal Pradesh and Gujarat. Green fodder yield is 60–95 t/ha.

16.19.3 LUCERNE NO. 9-L

This variety has been developed at Punjab Agricultural University, Ludhiana. It is a quick growing variety with deep green foliage, slender stalks, and purple flowers. It grows well for a period of 5–7 years. Its yield potential is about 75 t/ha of green fodder/year. It yields 57.7 t/ha of green fodder up to July during the first year.

16.19.4 LL COMPOSITE 5

Synthesized selecting 125 downy mildew-resistant clones from Kutch Lucerne at PAU, Ludhiana; released in 1981 for Punjab. It is tall, erect, fast-growing annual variety. It gives eight cuttings up to first week of July and it has a yield potential of about 72 t/ha fodder and 0.3–0.5 t/ha seed.

16.19.5 RAMBLER

It is a recent introduction from Canada and has been found successful in hilly areas of the country. It is tolerant to very low temperatures. Its yield potential is about 60–90 t of green fodder/ha/year.

16.19.6 LL COMPOSITE 3

It is synthesized from 20 clones selected for fast growth, high yield, and downy mildew resistant from germplasm collected from Gujarat, released in 1985 for entire country. It is resistant to lodging and frost with 39 t/ha green fodder yield in rabi season and 0.32 t/ha seed yield.

16.19.7 CHETAK (S-244)

A selection from local material of Maharashtra. Suitable for Punjab, Haryana, Uttar Pradesh, and Gujarat. It has quick regeneration capacity with resistance to aphids. It yields 142 t/ha green fodder.

16.19.8 RL 87-1, RL 88

Suitable for Maharashtra, Madhya Pradesh, and Uttar Pradesh. Green fodder yield is 80–95 t/ha.

16.19.9 NDRI SELECTION NO. 1

It has thick roots which penetrate deep into soil. It has turgid stems. The leaves are smaller in size when compared to other lucerne varieties. This variety has the capacity of maintaining itself in its pure stands over 5–6 years without getting degenerated due to the infestation of weeds. The crop is ready for first cut after 60–70 days of sowing. Its green fodder yield potential is about 100 t/ha.

16.19.10 SIRSA TYPE 9

This perennial variety has also been developed at Fodder Research Station, Sirsa. It is a quick-growing variety with deep green foliage. Its yield potential is about 30–40 t/ha of green fodder and 0.25–0.43 t/ha seed. It is most suitable for growing in North India.

16.19.11 CO-1

Perennial lucerne cultivar suitable for Tamil Nadu and Karnataka. Green fodder yield is 60–80 t/ha.

16.19.12 T 9

Perennial lucerne cultivar, suitable for entire lucerne areas of country. Fodder yield is 80–95 t/ha.

16.19.13 ANAND-2 (GAUL-1)

A selection from perennial-type lucerne grown in Bhuj area of Kutch (Gujarat). Released in 1975, suitable for Gujarat, Rajasthan, and Madhya Pradesh. Yield 80–100 t/ha green fodder in 10–12 cuts/year and 0.2–0.3 t/ha of seed may be obtained.

Besides the above varieties, there are also some promising varieties such as Moopa, IGFRI S-54, IGFRI S-244, IFGRI 112 (suitable for all areas), Nimach 1, Nimach 2, Composite 3, T8, and T15.

16.20 CUTTING MANAGEMENT

The first cut should be taken at 55–60 days after sowing and the subsequent cuts may be taken at 25–30 days interval when crop attains the height of 60 cm from the surface of the soil. In general, perennial lucerne gives seven to eight cuts in entire season.

16.21 YIELD

The average green fodder yield of lucerne varies from 80–100 t/ha.

16.22 SEED PRODUCTION

Higher seed yields are obtained from plant crop, which is not cut for fodder. For seed production, cutting should be stopped at last week of January. Irrigation should be stopped after full blooming to arrest further vegetative growth, and thus ensures good seed yield. The seed crop should be sown in rows 50-cm apart. Foliar spray of 0.5% borax at preflowering stage is found promising for seed production. Harvesting is done prior to maturity of crop to avoid the shedding of pods. The seed yields usually vary from 2.0–3.0 quintals/ha.

16.23 ANTIQUALITY CONSTITUENT BLOAT

The risk of producing bloat in animals is very real when cattle are grazed. The methods of controlling the bloat have often been troublesome, expensive, and only partly effective. To reduce this problem, lucerne should be sown with grasses because feeding lucerne mixed with grasses does not cause bloat to the cattle. Bloat causes due to presence of certain glucosides known as saponins in lucerne plant at green stage. Raw unsprouted alfalfa has toxic effects in primates, including humans, which can result in lupus-like symptoms and other immunological diseases in susceptible individuals (Montanaro and Bardana, 1991).

16.24 LUCERNE SILAGE

Fermented high-moisture fodder that can be fed to ruminants is known as silage. Lucerne is high in protein and low in soluble carbohydrates which are needed to enable anaerobic bacteria to produce lactic, acetic, and propionic acids which preserve the forage as silage. These acids reduce the pH and inhibit further bacterial and enzyme action. Avoid exposure to air to reduce disappearance of available carbohydrate and deterioration of the plant material. *Clostridium*, an undesirable bacteria, grows under high pH and produces butyric acid which makes the forage very unpalatable. Additives are added to lucerne silage to reduce the bad smell as well as allowing the crop to wilt. Care should be taken when handling the wilted material to avoid loss of leaf. Moisture content should be less than 65% and no free juice apparent for successful silage making.

16.25 INSECTS

The main pests of lucerne at establishment are red-legged earth mite and lucerne flea. Other insects include aphids, weevils, and lucerne caterpillar. Being a leguminous and perennial forage crop, insect-pests attack is a common phenomenon. Insects are likely to invade from previous stand of lucerne. Aphid damage appears as leaf curling and stunted plants. Small lucerne weevil fed on leaves leaving crescent-shaped chew marks in leaves. Monitor regularly as damage can occur before mites can be seen. Spray miticides during postsowing preemergent to control mites. Check closely for aphid damage in the first spring and spray any systemic insecticide as necessary. Spray of 0.2% carbaryl (Sevin) reduces the infestation of caterpillar (Singh, 2005).

16.26 DISEASES

Bacterial wilt is the most serious disease of lucerne caused by *Aplanobacter insidiosum*. Infected plants get stunted and many branched. Resistance cultivars should be grown to control this disease. Another disease is leaf spot which is caused by *Pseudopeziza medicaginis*. Diseased plants turn yellow and leaves drop off. Early cutting can reduce the infestation to some extent or 0.2% spraying of Mancozeb 75 WP controls this disease.

FIGURE 16.1 (See color insert.) Lucerne.

KEYWORDS

- **lucerne**
- **agronomic management**
- **fodder yield**
- **nutritive value**
- **antinutritional factors**
- **utilization**

REFERENCES

Montanaro, A.; Bardana Jr, E. J. Dietary Amino Acid-induced Systemic Lupus Erythematosus. *Rheum. Dis. Clin. North Am.* **1991,** *17* (2), 323–332.

Singh, S. S. *Crop Management*, 5th ed.; Kalyani Publishers: New Delhi, India, 2005.

Singh, C.; Singh, P.; Singh, R. *Modern Techniques of Raising Field Crops*, 2nd ed.; Oxford and IBH Publishing: New Delhi, India, 2011.

CHAPTER 17

STYLO (PENCILFLOWER)

PARTHA SARATHI PATRA* and TARUN PAUL

Department of Agronomy, Uttar Banga Krishi Viswavidyalaya, Pundibari 736165, Cooch Behar, West Bengal, India

Corresponding author. E-mail: parthaagro@gmail.com

ABSTRACT

Stylo is an important pastures and forage species. It is a tropical legume shrub or herb widely grown for forage throughout the tropics and subtropics. It is drought tolerant and can be harvested during the whole year, though it grows only when sufficient water is available. It contains 12–18% crude protein, 0.61–1.72% ca., 0.10–0.12% phosphorus, and 7.0–14.2% ash. It often needs supplementary phosphorus to achieve best animal performance. Sodium levels are much higher than those for many other tropical legumes and may be 1–2% of dry matter in leaf and stem. The green forage production ranges from 200 to 300 quintals/ha, while the dry forage is 100–200 quintals/ha depending on the soil fertility. It generally produces 350–400 kg seeds/ha but from well-managed pasture, seed production reaches up to 1000 kg/ha.

17.1 BOTANICAL CLASSIFICATION

Kingdom: Plantae
Order: Fabales
Family: Fabaceae
Subfamily: Faboideae
Tribe: Aeschynomeneae
Subtribe: Stylosanthinae

Genus: *Stylosanthes*
Species: *hamata* (L.)
Binomial name: *Stylosanthes hamata* (L.)

17.2 COMMON NAME

Shrubby stylo (Australia), capitan juan, pata de terecay (Venezuela), pencilflower (United States), alfafa do nordeste (Brazil), stylo perduan (Indonesia)

17.3 BOTANICAL NAME

Stylosanthes hamata (L.)

17.4 INTRODUCTION

Stylosanthes is a genus of flowering plants in the legume family Fabaceae and contains numerous highly important pasture and forage species. It was recently assigned to the informal monophyletic *Pterocarpus* clade of the *Dalbergieae*. It is a tropical legume shrub or herb widely grown for forage throughout the tropics and subtropics. It is drought tolerant and can be harvested during the whole year, though it grows only when sufficient water is available. It contains 12–18% crude protein, 0.61–1.72% ca., 0.10–0.12% phosphorus, and 7.0–14.2% ash. It often needs supplementary phosphorus to achieve best animal performance. Sodium levels are much higher than those for many other tropical legumes and may be 1–2 % of dry matter in leaf and stem. Being nutritive and palatable, it is used as feed for all types of animals in the form of hay, silage. Among the species of *Stylosanthes*, *Stylosanthes hamata* and *Stylosanthes viscosa* lines were found to be rich in crude protein and lower in fiber contents as compared to *Stylosanthes scabra* and *Stylosanthes sebrana* lines. Stylo is not often sown with other legumes but it can be intercropped with rice, maize, or cassava, depending on soil fertility. Stylo can be grazed but it is sensitive to heavy grazing. It should not be grazed until 6–8 weeks after sowing. Rotational grazing is preferable with 4–8-week rest intervals.

17.5 ECONOMIC IMPORTANCE

Species within the genus have many properties that make them valuable forage species; it can be mixed with tropical grasses such as *Brachiaria* spp., *Andropogon gayanus*, *Chloris gayana*, *Digitaria eriantha*, *Heteropogon contortus*, *Hyparrhenia rufa*, *Melinis minutiflora*, *Pennisetum purpureum*, or *Setaria sphacelata*. It has acaricidal properties, with the lethal effects on larval and nymphal ticks of *Rhipicephalus sanguineus*, *Boophilus microplus,* and *Haemaphysalis intermedia.*

Stylo can be easily cut and then fed fresh to livestock. It is, however, not very palatable when young and it is advised to wilt it to soften its bristles before offering it to the animals and it is a valuable deferred feed for cattle as its palatability increases with maturity. Stylo can make valuable hay but should be handled carefully so that it does not shed its leaves. For sward longevity, stylo should not be cut below 20 cm and no more than once a year. Stylo may be used as silage when ensiled with salts and molasses.

Stylo has been used to improve the nutritive value of natural grasslands in Australia. Due to heavy and fast leaf fall, it helps in improving the soil and is also used for checking the soil erosion. Stylo is an N-fixing legume that readily nodulates and improves soil N mineral status. It is able to extract P from soils that are very poor in this nutrient and it is tolerant of low Mo levels. In Laos, stylo fallow increased rice yield and decreased weed biomass. In Nigeria, a stylo fallow preceding a maize crop resulted in a yield of maize similar to that obtained with the addition of 45 kg N/ha. Stylo was reported to control weeds such as *Striga asiatica*, *Rottboellia exaltata*, *Borreria alata*, *Boerhavia diffusa,* and *Imperata cylindrica*. It is a valuable cover crop in coconut and palm oil plantations. While normally used for ruminant production, *Stylosanthes guianensis* is also used to feed pigs in Southeast Asia.

17.6 BOTANICAL DESCRIPTION

Taxonomy of the genus remains unsettled and controversial, with various authors favoring between 25 and 42 species, with at least 40 additional synonyms. The taxonomy is complicated by the existence of numerous natural tetraploid and hybrid populations. Species within the genus fall within two subgenera: Styposanthes and Stylosanthes. Styposanthes possess a small rudimentary secondary floral axis which is absent in Stylosanthes. Stylosanthes is closely related to the peanut genus Arachis.

Stylo is an annual or perennial to short-lived, erect or semi-erect shrubby or herbaceous non-determinate legume that can reach a height of 75–2.0 m depending on the species. Stylo has a strong deep taproot (goes up to 4-m depth of soil) that is nodulated. Stems spreading to ascending, with a dichotomously branching habit, smooth and may be woody at the base. Young stems vary from green to reddish in color, depending on the strain; usually with dense hairs and bristles, and viscid, becoming woodier with age. Stylo does not twin, unlike other legumes. Stylo is a leafy species that remains green under dry conditions.

The leaves are trifoliolate with elliptical to oblong-lanceolate leaflets, pale green to dark green and dark blue-green in color; terminal leaflet 20–33-mm long, 4–12-mm wide.

The inflorescence is a densely flowered spike, 1–3-cm long, with up to 40 flowers/head. Flowers are yellow to orange with black or red stripes. The fruit is a one-seeded pod, 2–3 mm long by 1.5–2.5-mm wide. The pods or so-called seeds are medium to dark brown in color, 2–2.5-mm long, asymmetrical by reniform, radical ends fairly prominent, and beak is slightly coiled. Actual seed comes after removing the brown covering and is light yellow or pale brown or purple in color depending on the species.

Two major types (Brazilian coastal and Continental type) are recognized in northern South America as follows:

1. Brazilian coastal type from higher rainfall districts close to coast—late flowering, taller, more erect, higher yielding, often most anthracnose resistant.
2. Continental type from Colombia, Venezuela, and central Brazil—early flowering, shorter, semi-erect, less vigorous, more anthracnose susceptible; often reddish stems.

Accessions collected in northern Argentina and identified as S. scabra, are low growing, fine stemmed, and early flowering (January/February at 17°S), bearing a stronger resemblance to *S. guianensis* var. *intermedia* than to the shrubby types of S. scabra of northern South America.

17.7 ORIGIN AND DISTRIBUTION

All except two species of the genus are native to the Americas. S*tylosanthes* fruticosa has a native range that extends from South Africa to Ethiopia, across Arabian, Peninsula to Pakistan, India and Sri Lanka and S*tylos-anthes* erecta is endemic to Tropical Africa, from Tanzania to Senegal. The putative species Stylosanthes sundaica, has a range that encompasses Malaysia but is considered by most authors to be an adventives polypoliod variety of *Stylosanthes humilis*. Ecological range extends from savanna and thorn scrub to tropical forest and montane forests.

S. hamata occurs mainly in the West Indies, in the Caribbean area generally and in coastal areas of Venezuela and Colombia bordering the Caribbean. It also occurs in coastal areas of southern Florida, United States. It is occasionally reported from Honduras and various parts of Brazil but these may not be native populations, especially those growing in isolated areas in southern Brazil.

S. scabra is a promising legume which is performing well under South Indian Agro-climatic region.

17.8 ECOLOGICAL REQUIREMENT

Stylo is found from 20°N to 32°S, and from sea level up to an altitude of 2000 m (Mannetje, 1992). Stylo can grow in places where annual rainfall ranges from 700 to 5000 mm, but it does better between 1000 and 2500 mm for common stylo and between 600 and 1800 mm for fine stem stylo. Stylo is a warm season growing legume that thrives in places where annual temperatures are between 23°C and 27°C. However, stylo can survive light frost (0°C) and can remain productive down to 15°C. Fine stem stylo has more frost tolerance than common stylo.

Primarily a short-day flowering response, with critical photoperiod lies between 11.5 and 12.5 h depending on ecotype. In northern Australia, flowering dates vary from mid-January to mid-May.

17.9 SOIL

Stylo does well in most soils from sands to light clays (including those that are relatively infertile or deprived of P) provided they are well-drained.

Soil pH ranging from 4 to 8.3 is acceptable to var. *Guianensis,* which also has some tolerance of aluminum and manganese. Fine stem stylo prefers neutral soils. Heavy clay soil is not suitable for its cultivation. It cannot tolerate salt.

17.10 ESTABLISHMENT

Plow the field two to three times to obtain good tilth. Stylo can be sown alone or mixed with companion species. In Australia, it is often over sown in native grasslands. Stylo can be sown in plots (7–12 seeds/plot) and should not be buried as the seeds are very small. Stylo can be broadcast when over seeded in grassland. In humid areas, stylo can be sown at any time provided that there is no dry period during its establishment. In drier parts, it should be sown as soon as possible after the start of the rainy season, and at least 2 months before the rain stops (Husson et al., 2008). For pure pasture, the recommended seed rate is 12 kg/ha but for mixed pasture, the seed rate is 6 kg/ha. Seeds are sown in July after first heavy shower either in line at 50-cm space or broadcast. Sowing can be done up to October. Sowing depth is 1.0–1.5 cm (Saito et al., 2006).

17.11 SEED TREATMENT

Stylo seeds possess hard seed coat. So, acid scarification is to be done by dipping the seeds in concentrated sulfuric acid for 3 min and washing thoroughly with tap water. Acid scarified seeds are again to be presoaked in cold water overnight. Seeds can also be scarified in hot water. Before sowing, the seeds should be scarified or treated with hot water for 1–1.5 min. Seeds are to be treated with cowpea group rhizobium culture for better nodulation.

17.12 NUTRIENT MANAGEMENT

Apply nutrient as per soil test recommendation as far as possible. If the soil testing is not done, apply 5–8 t FYM/ha at the time of land preparation and 20 kg N and 40–60 kg P_2O_5/ha and 40 kg K_2O/ha before sowing. From second year onwards, 30 kg P_2O_5 and 15 kg N/ha are sufficient. Basal application of NPK is advocated for better yield.

17.13 WATER MANAGEMENT

It is a rainfed crop. But during the period of establishment, care should be taken to provide sufficient moisture.

17.14 WEED MANAGEMENT

Hand weeding may be given as and when necessary. During the establishment year, 1–2 weeding and inter-culturing are required for better growth. Herbicides are largely used for seed production only. Trifluralin and 2,4-D are safe and effective in establishment.

17.15 PESTS AND DISEASES

Variable susceptibility to anthracnose caused by *Colletotrichum gloeosporioides* and *Colletotrichum dematium*—controlled by use of resistant varieties. Head blight caused by *Botrytis cinerea*, can be a problem in seed crops in years with overcast weather during flowering. Sclerotium blight caused by *Corticium* (*Sclerotium*) *rolfsii* and sclerotinia stem rot caused by *Sclerotinia sclerotiorum* can kill plants when environmental conditions favor their development.

A disease often referred to as "reversion," (reverts from reproductive to vegetative state) caused by a phytoplasma has become a major problem in seed crops. It occurs late in the season, producing elongation of inflorescences and interfering with seed set, ultimately killing plants. It is carried by a leafhopper (*Erosius* sp.) and affects a range of other species in the seed production area. The condition also occurs in pasture plantings but is much less severe.

Violet root rot caused by *Rhizoctonia crocorum* (teleomorph *Helicobasidium*) has led to stand reduction of epidemic proportions in the sub-humid subtropics during cool, wet conditions. Stem borers, the larvae of *Caloptilia* sp. (Lepidoptera: Gracillariidae) in Brazil and Colombia, and *Platyomopsis pedicornis* (Coleoptera: Cerambycidae) can cause significant damage in pastures and seed crops.

17.16 VARIETY

Several varieties of stylo have been evolved for cultivation in different countries. The brief description of some of the important cultivars is listed below in Table 17.1. In India, little work had been reported so far for the development of new varieties; however *Stylosanthes scabra*, released from TNAU and RS 95 released from MPKV, Rahuri has been found promising for southern region and western region respectively. During 1970s, a major outbreak of the disease anthracnose, caused by the fungus *C. gloeosporioides*, devastated stylo cultivation throughout the world, and popular cultivars such as Schofield, Cook, and Graham were found highly sensitive to this disease. Since then, breeding efforts have focused on developing anthracnose-resistant cultivars. CIAT 184, developed in Peru, is resistant to anthracnose in the humid tropics, and cultivars derived from this line have been successfully cultivated in South America, China, Southeast Asia, Vietnam, Thailand, and the Congo (Kinshasa). Other varieties of *Stylosanthes guianensis* are also available as commercial cultivars. Ubon stylo (*Stylosanthes guianensis* var. *vulgaris* x var. *pauciflora*) is anthracnose-resistant. Fine stem stylo has a lower drought tolerance than common stylo but a higher resistance to anthracnose. One fine stem stylo cultivar (Oxley), bred in Australia at the end of the 1960s is adapted to sandy soils of the subtropics.

17.17 HARVESTING

First harvest can be taken 75 days after sowing at flowering stage and subsequent harvests depending upon the growth. During establishment year (first year), it should not be allowed to be grazed at all but should be harvested at the height of 10 cm from ground level after four months of sowing. From second year onwards, it may be grazed or harvested two to three times. Rotational grazing is preferred for higher production.

17.18 FODDER YIELD

The green forage production ranges from 200–300 quintals/ha, while the dry forage is 100–200 quintals/ha depending on the soil fertility. It

TABLE 17.1 Brief Descriptions of Some of the Important Cultivars.

Sr. no.	Cultivars	Country/date of release	Details
1.	"Feira"	Australia (1990)	F5 selection of the cross, Q10042 × CPI 55860. Resistant to races 1, 3, and 4 of *Colletotrichum gloeosporioides* causing Type A anthracnose disease. Shortest of "Siran" components. Flowers earlier than "Seca" but later than "Recife"; has more tertiary branches than "Fitzroy" and "Seca"
2.	"Fitzroy" (CPI 40205)	Australia (1980)	From Cruz Das Almas, Bahia, in eastern Brazil (12°S, 36 MSL, rainfall ca. 1150 mm). Vigorous, erect, somewhat spreading plant, 1.5–2 m tall, with green stems. Selected as an earlier flowering, leafier alternative to "Seca" for use in the subhumid subtropics. Proved to be susceptible to type A anthracnose. Now rarely planted, even in drier areas where anthracnose is not a problem, because commercial seed is not available
3.	"Jecuipe"	Australia (1990)	F5 selection of the cross Q10042 × CPI 93116. Resistant to races 1, 3, and 4 of *Colletotrichum gloeosporioides* causing Type A anthracnose disease. Flowers earlier than "Seca" and slightly later than "Recife." Plants are shorter than those of "Seca" or "Recife." Has denser tertiary branching than "Seca" or "Fitzroy"
4.	"Recife"	Australia (1990)	Selection from genetically variable population of "Seca." Resistant to races 1 and 4, and moderately resistant to race 3 of *Colletotrichum gloeosporioides*, which cause Type A anthracnose disease. Early flowering and intermediate height, plants being taller than those of "Jecuipe" and "Feira" but shorter than of "Seca"
5.	"Seca" (CPI 40292)	Australia (1977)	From Pernambuco, northeast Brazil (8.13°S, 35.30°W, 300 m asl, rainfall 800 mm, 7-month dry season). Vigorous, erect plant to 2-m tall, with reddish stems and open canopy. Widely adapted. Single gene resistance to anthracnose
6.	"Siran"	Australia (1990)	Composite of three bred or selected lines ("Jecuipe," "Recife," "Feira") to provide multiple-gene resistance to anthracnose. Generally similar in appearance and adaptation to "Seca"

generally produces 350–400 kg seeds/ha but from well-managed pasture, seed production reaches up to 1000 kg/ha.

17.19 SEED PRODUCTION

Seed can be hand or machine harvested. Suction harvesting is possible, but rarely economical. Because of the span of flowering time, peak seed yields can span a period of 3–4 weeks. For machine harvest, seed recovery is favored by using a machine with a small front, high horse-power, slow groundspeed, and large sieve area. Also best to harvest during low-humidity conditions when seed is less sticky, and when seed in the head has not been dislodged by recent strong winds. In northern Australia, seed is normally harvested during the dry season. Irrigated crops are harvested in August–September, and rain grown crops, a month earlier. In the past, seed stands were retained for as long as possible, but needed renovation after 2–4 years, as older plants became moribund. Seed crops are now treated as annuals due to the emergence of a phytoplasma disease (often referred to as "reversion"), which tends to attack plants in the second year after sowing. Seed is sown early each season and the crop irrigated to ensure rapid development and earlier floral initiation. If harvesting is delayed beyond the cooler part of the dry season, incidence of "reversion" increases. Trash is destroyed by burning or incorporation into the soil following harvest to reduce phytoplasma in the system. Yields of seed-in-pod vary from 100–700 kg/ha, mostly 300–400 kg/ha.

KEYWORDS

- **stylo**
- **agronomic management**
- **fodder yield**
- **nutritive value**
- **utilization**

REFERENCES

Husson, O.; Charpentier, H., Razanamparany, C.; Moussa, N.; Michellon, R.; Naudin, K.; Razafintsalama, H.; Rakotoarinivo, C.; Rakotondramanana and Séguy, L. *Stylosanthes Guianensis*. Manuel Pratique Du Semis Direct À Madagascar, Volume III. Fiches Techniques Plantes De Couverture: Légumineuses Pérennes. Madagascar, 2008.

Mannetje, L. T. *Stylosanthes guianensis* (Aublet) Swartz. Record from Proseabase. Mannetje, L.'t and Jones, R. M. Eds. PROSEA (Plant Resources of South-East Asia) Foundation, Bogor, Indonesia, 1992.

Saito, K.; Linquist, B.; Keobualapha, B.; Phanthaboon, K.; Shiraiwa, T.; Horie, T. 2006. *Stylosanthes guianensis* as a Short-term Fallow Crop for Improving Upland Rice Productivity in Northern Laos. Field Crop Res. **2006,** 9 6 (2–3), 438–447.

CHAPTER 18

PESTS OF LEGUMINOUS FORAGES

N. NAIR[1*], B. C. THANGJAM[1], UTPAL GIRI[1], and M. R. DEBNATH[2]

[1]*Department of Entomology, College of Agriculture, Lembucherra 799210, West Tripura, Tripura, India*

[2]*Horticulture Research Centre, Nagicherra, West Tripura 799004, Tripura, India*

[*]*Corresponding author. E-mail: navendunair@gmail.com*

ABSTRACT

Leguminous forage crops harbor several insect pests of which some are responsible for economic damage to the crops. The present chapter presents an overview of pest problems in leguminous forage crops and their management. In this chapter, pest spectrum, pest descriptions, damage symptoms, and management of common pests of forage legumes of various groups, namely, annual (cowpea, soybean, berseem, cluster bean, fenugreek, velvet bean, and moth bean), perennial forage (lucerne and stylos), and perennial trees (agathi, subabul, *Gliricidia sepium*, and *Prosopis* spp.) have been given. The pest spectrum provides the information about various insect pests occurring in a particular crop. Pest descriptions of major pests and some other minor but common pests have been included. The methods of pest management are broadly based on the current state of general acceptance and use and also from available literatures.

18.1 INTRODUCTION

The pests always exist simultaneously with the cultivated crops in the agro ecosystem. Though the insect and other pests play an important role in maintaining the ecological balance, their presence may sometimes cause economic damage to the crops. Moreover, the pests are considered major or minor based on nature and extent of damages caused by them. Leguminous forage crops harbor a number of insect pests. Both direct and indirect losses are caused by these pests. Impairment of forage quality and reduction of green fodder yield are considered as the direct losses. Indirect losses may be addressed to the reduction of nodule formation in roots and in turn reduction of nitrogen fixation capacity. Some sucking pests such as aphids and white flies are also responsible for the transmission of viral diseases in addition to direct losses to the crop by desaping. Since the insect pests restrict the production of forage legumes there is considerable scope for increasing the production by controlling the pests. Proper identification of the pest species and their potentiality of causing economic loss are to be ascertained before formulating pest management strategies. Hence, regular monitoring is the prime requirement of a successful pest management program. However, though chemical methods of pest control are known and a lot of chemical pesticides are available in the market; integrated approach of pest management including the development of pest resistant cultivars should be emphasized.

18.2 LEGUMINOUS ANNUAL FORAGES

18.2.1 PESTS OF COWPEA (Vigna unguiculata (L.) WALP.)

The losses in grain or foliage of cowpea ranges from 20% to 100% due to field insect pests (Raheja, 1976; Singh and Allen, 1980). The avoidable losses in yield due to insect pests have been recorded in the range of 66–100% in cowpea (Pandey et al., 1991). The important insect species attacking cowpea crop include aphid (*Aphis craccivora* Koch), leafhopper (*Empoasca kerri* Pruthi), thrips (*Megaleurothrips distalis* Karny), whitefly (*Bemisia tabaci*, Genn.), leaf miner (*Acrocercops caerulea* Meyrick), spotted pod borer (*Maruca vitrata* Fab.), tobacco leaf eating caterpillar (*Spodoptera litura* Fab.), and blue butterfly (*Euchrysops cnejus* Fabricius) (Patel et al., 2010). The pod borer complex posing serious threat to cowpea cultivation includes *Maruca vitrata* (Fabricius), *Lampides boeticus* (L.), *Helicoverpa armigera* (Hubner), *Etiella zinckenella* (Treitschke), *Adisura atkinsoni* (Moore), and *Exelastis atomosa* (Walsingham) (Subhasree and Mathew, 2014). Ram et al. (1989) studied the seasonal incidence of *Empoasca kerri* and defoliating pests (*Pagria signata, Plusia nigrisigna*

[*Autographa nigrisigna*], *Spodoptera litura* Fab., *Colemania sphenarioides*, *Chrotogonus trachypterus*, and *Atractomorpha crenulata*) on fodder cowpeas (*Vigna unguiculata*) in the Bundelkhand region of Jhansi, India.

18.2.1.1 BLACK APHID: Aphis craccivora Koch (HEMIPTERA: APHIDIDAE)

A. craccivora is polyphagous, but with marked preference for Leguminous host plants. It is a major economic pest of groundnut, cowpea, mungbean, pigeon pea, chickpea, beans, lentil, and lucerne. This aphid is relatively small in size (1.4–2.2 mm) and shiny black or dark brown in color. The immature stages are lightly dusted with wax and pass through four (range 3–5) moults.

On cowpea, aphids normally feed on the under surface of young leaves, tender shoot, and green pods of mature plants. Severe infestation results in stunted growth, distorted leaves, premature defoliation, and death of seedlings. This aphid is known to transmit cowpea aphid-borne mosaic virus.

The damage to the crop results in profuse draining of plant sap and development of honeydew leading to black sooty mould on leaves and leaf shedding (Kotadia and Bhalani, 1992).

Management

Avoid using heavy doses of nitrogenous fertilizers. Spray a steady stream of water on the host plant to knock-off aphids. Use yellow sticky traps to attract the alate adults. Parasites and predators especially Coccinellids, Syrphids, and Chrysopids reduce the population of aphids considerably. Release of *Menochilus sexmaculata* @ 1250/ha or *Chrysoperla carnea* grubs @ 5000/ha. Apply neem seed kernel extract 5%. Application of Imidacloprid 17.8 SL. @ 200 mL or Thiamethoxam 25 WG @ 100 g or Acephate 75 SP. @ 500 g or dimethoate 30 EC @ 650 mL/ha in 500 L of water per hectare.

18.2.1.2 WHITEFLY: Bemisia tabaci Gennadius (HEMIPTERA: ALEYRODIDAE)

The whitefly, *B. tabaci* is widely distributed and highly polyphagous insect pest and is known to feed on several vegetables, field crops, and weeds.

The nymphs are flattened, oval-shaped, and greenish-yellow in color. Last three instars nymphs are immobile and look like scales. The last nymphal stage has red eyes and it is often referred to as pseudo pupa. Adult whitefly is minute insect, about 1-mm long, soft-bodied, light yellow in color. The wings are covered with powder like waxy materials.

This insect feeds on the plants by sucking sap from the underside of leaves. As a result of feeding by large numbers of whiteflies in case of severe infestation, the entire leaf turns yellow, chlorophyll of leaves is destroyed, plants growth is stopped, and the loss may range from 12% to 65% (Kumar and Narain, 2005). The insect secretes honeydew on which growth of sooty mould takes place resulting in blackening of leaves, drastically reducing photosynthetic rate and drying of leaves leading to total failure of the crop. *B. tabaci* is also of considerable importance because it also transmits the viral diseases in cowpea. One of the most important viruses infecting cowpea is cowpea golden mosaic virus which is actively transmitted by *B. tabaci* (Shaonpius and Charanjit, 2010).

Management

Grow maize, sorghum or pearl millet as a barrier crop to minimize the incidence of whiteflies. Grow cotton as a trap crop 1 month earlier between the cowpea rows. Use yellow sticky traps at the rate of 1–2 traps/50–100 m^2 to trap adult whiteflies. Foliar spray of triazophos 40 EC @ 500 mL or Imidacloprid 17.8 S.L. @ 200 mL or Thiamethoxam 25 WG @ 100 g or Acephate 75 S.P. @ 500 g in 500 L of water per hectare is effective in controlling whitefly.

18.2.1.3 LEAF HOPPERS OR JASSID: Empoasca kerri Pruthi (HEMIPTERA: CICADELLIDAE)

The jassids or leafhoppers are widely distributed and one of the destructive pests in many regions of the country. It is potential pest of cowpea.

The jassids are wedge shaped and walk diagonally. Adults are about 3 mm in length and are of greenish in color. Both the winged adults and nymphs are green in color and suck the cell sap from tender leaves, twigs and flowering, and fruiting bodies of the crop. Besides damaging the plant through desapping; they also inject some toxins in the plants. The symptoms of damage are yellow discoloration ofthe leaf veins and margins, followed by cupping of the leaves. Severely infested plants become stunted and may dry prematurely. It also secretes honey due to which leads to development of sooty moulds on the leaf surface affecting photosynthesis, growth, and yield of the crop. *E. kerri* causes yield reduction up to 39% (Singh and Van Emden, 1975).

Management

Quinalphos, carbaryl, and dimethoate may provide the crop good protection up to 20–25 days. Application of seed coating of aldicarb @ 1.0 kg a.i./ha has also been highly effective. In addition to use of aldicarb at sowing time, monocrotophos spray in 3–5-week-old cowpea crop, has also been effectively recommended for control of Jassids in cowpea (Kumar and Narain, 2005).

18.2.1.4 POD BUG: Riptortus linearis Linnaeus AND R. pedestris Fabricius (HEMIPTERA: ALYDIDAE)

Riptortus spp. are very common pest of pigeon pea, lablab, cowpea, and other pulses. It is generally a minor pest but sporadically may be serious.

The adult bug is cylindrical, light brown with characteristic white or yellow lines on the side of the body in case of *R. linearis*, and yellow dots on the side of the thoracic segments in case of *R. pedestris*. They suck the sap from green pods showing feeding punctures inside and grains become shrivelled. Tender pods when attacked, fail to develop fully. The management is same with the green stink bug and is given for both as management of pod sucking bugs.

18.2.1.5 GREEN STINK BUG: Nezara viridula Linnaeus (HEMIPTERA: PENTATOMIDAE)

It is primarily a major pest of soybean but sometimes causes extensive damage to cowpea crops. Nymphs are shiny with bright spots, whereas adults are green and triangular in shape. Both adults and nymphs suck sap from green pods, resulting in shriveled and prematurely dried pods and seeds.

Management of Pod Sucking Bugs

Collection of bugs and their destruction by dipping into kerosinized water and dusting with carbaryl 10D 10 kg/ac or foliar spray with dimethoate 30 EC @ 2 mL/L or monocrotophos 36 SL @ 1.5 mL/L of water are effective.

18.2.1.6 BLISTER BEETLE: Mylabris pustulata Thunberg (COLEOPTERA: MELOIDAE)

Blister beetle is highly polyphagous and it feeds on the flowers of several plants in the families: Convolvulaceae, Cucurbitaceae, Leguminosae, Malvaceae, etc. A total of 52 host plants belonging to 25 botanical families were found to be attractive to *M. pustulata* (Durairaj and Ganapathy, 2003).

The adult beetle is about 2.0–2.5 cm in length, stout bodied and elongated in shape with broad red or reddish-orange and black alternating bands on the forewing or elytra. The grubs undergo hypermetamorphosis, with the different larval instars of different forms. Grubs are predatory, remain under soil and do not feed on plant material. Adult beetles are voracious feeder and cause severe damage to buds, flowers, tender pods, and even tender leaves resulting in reduced yields.

Management

Adults can be collected by using insect-catching nets. *For controlling this pest application of carbaryl 5% dust @ 20–25 kg/ha is recommended.* Synthetic pyrethroids may be used for a quick knock-down effect.

18.2.1.7 SPOTTED POD BORER: Maruca vitrata (FABRICIUS) = (Maruca testulalis Geyer) (LEPIDOPTERA: CRAMBIDAE)

It is commonly found in association with many legumes such as pigeon pea, field bean, cowpea, green gram, black gram, soybean, etc. and considered as a serious pest on these crops.

Larval body is semitransparent and spotted on each segment and the spotting intensity varies and the spots fade before pupation. The pupae are elongated, measuring about 13 mm in length and with shouldered appearance. Early pupal stage is greenish but turns brown when fully developed and concealed in a cocoon on dry leaves, flowers, and other dead plant matters. The Adult are medium sized and both sexes are morphologically alike. The forewings are brown having white spot and black-edged while the hind wings are semi-hyaline (Ashigar and Umar, 2016).

It is a very common and occasionally serious pest of cowpea. The larvae damage flower buds, flowers, and developing pods and also web the inflorescences. Kumar et al. (2013) recorded 22.8–32.56% pod damage among the test cultivars of cowpea by this pod borer pest.

Management

Foliar spray from flower bud initiation with chlorpyriphos 20 EC @ 2.5 mL/L or quinalphos 25 EC @ 2 mL/L or novaluron 10 EC @ 0.75 mL/L or spinosad 45 SC @ 0.50 mL/L or lamda cyhalothrin 4.9 CS @ 1 mL/L or flubendiamide 39.35% SC @ 0.25 g/L or dichlorvos 76 EC @ 1 mL/L of water at weekly intervals is effective.

18.2.1.8 HAIRY CATERPILLAR: Somena scintillans Walker (LEPIDOPTERA: EREBIDAE)

It is a highly polyphagous pest and commonly found on ragi, castor, pigeon pea, cowpea, field bean, cucurbits, mango, rose, etc.

Generally it is a minor pest but sporadically it may be serious on cowpea. Flowers of cowpea are mostly consumed by this pest as compared to other plant parts.

18.2.1.9 LABLAB LEAF WEBBER: Omiodes indicata Fabricius (LEPIDOPTERA: CRAMBIDAE)

It is mostly associated with legumes and host range includes pigeon pea, lablab, groundnut, cowpea, black gram, green gram, soybean, lucerne, etc.

Caterpillars are greenish in color. Moths are straw-colored with wavy transverse lines on the dorsal side of fore wings. The larvae web together the leaves and feed from inside by scraping of leaf tissues resulting in skeletonization and drying up of infested leaves. It is generally a minor pest and does not warrant any control measure.

18.2.1.10 PEA BLUE BUTTERFLY OR LONG TAILED BLUE: Lampides boeticus Linnaeus (LEPIDOPTERA: LYCAENIDAE)

It is commonly found on pigeon pea, field bean, cowpea, lablab, peas, and several other leguminous hosts.

Adults are having wings which are bluish on the upper side while the underside is pale brown with narrow whitish bands with a wingspan of about 3 cm. Hind wings have one or two black spots and a hair-like projection. Larvae are green, oval, flat, about 1-cm long and often attended by black ants. Larvae preferably feed on buds, flowers, and seeds within the pods. They get entry into the pod by making round boreholes and feed on developing seeds. Frass is deposited inside the pod causing decay and showing external dark discoloration.

Other closely related species are *Euchrysops cnejus* (Fabricius) commonly known as "gram blue" and *Catochrysops strabo* (Fabricius) commonly known as "Forget-me-not." These two species also occur on the same host plants with similar type of mode of damage.

Management

Spraying with crude neem extract (5%) or neem oil or carbaryl 50 WP @ 2.0 g/L of water.

18.2.2 PESTS OF SOYBEAN (Glycine max L. Merrill)

All stages of this crop are prone to heavy infestation by pest complexes. Some common insect-pests infesting soybean crops at various stages are Aphid (*Aphis gossypii* Glover), whitefly (*Bemisia tabaci* Gennadiu), jassids (*Empoasca kerri* Pruthi), leaf eating weevil (*Myllocerus* sp.), leaf miner (*Aproaerema modicella* Deventer), tobacco caterpillar (*Spodoptera litura* Fab.), American bollworm or gram caterpillar (*Helicoverpa armigera* Hub.), Bihar hairy caterpillar (*Spilosoma obliqua* Walker), leaf roller (*Omiodes indicata* Fab), stem fly (*Melanagromyza sojae* Zehntner, and *Ophiomyia phaseoli* Tryon), *Cirtocanthacris ranacea*, girdle beetle (*Obereopsis brevis* Swedenbord), green semilooper (*Chrysodeixis acuta* Walker), etc. (Gangrade, 1976; Rai and Patel, 1990; Uttam, et al., 2012; Naik et al., 2013; Biswas, 2013; Ahirwar et al., 2015).

18.2.2.1 SOYBEAN STEM FLY: *Melanagromyza sojae Zehntner* (DIPTERA: AGROMYZIDAE)

M. sojae is a common pest species of many leguminous crops including soybean, lucerne, cow pea, black gram, green gram, pigeon pea, etc. It is a major pest of soybean in India causing seedling mortality. Adults are small black flies. Maggots are apodous and yellowish in color. The adult stem fly deposit eggs in the leaf tissue of soybean seedlings. Maggots mine the leaves or bore into the leaf petiole or tender stem and cause extensive tunneling resulting in withering, drooping, and death of plant. The stem fly (*Melanagromyza sojae* Zehntner) caused maximum damage of 36.32%. The early stage maggots bore into the soybean stem and fed on to internal contents resulting in the death of seedlings (Naik et al., 2013).

Management

Since infestation by *M. sojae* at the seedling stage causes economic yield loss, it is essential to control this pest during the first 4–5 weeks after seed germination. It is important to know the season when the pest is serious to undertake appropriate prophylactic control measures. Precautionary control measures are to be taken in the dry season as the pest is more serious in the dry season than in the rainy wet season. Since larval damage is internal, control measures should necessarily be taken either at sowing (seed treatment) or immediately after germination (foliar spraying). Seed treatment with thiamethoxam 70 WS @ 3.0 g/kg seed and imidacloprid 70 WS @ 3 g/kg seed is very effective. Foliar spray with acephate 1.5 g/L or dimethoate 2 mL/L or monocrotophos 1.6 mL/L of water is effective.

18.2.2.2 SHOOT FLY: *Ophiomyia phaseoli Tryon* (DIPTERA: AGROMYZIDAE)

It is another common pest of many leguminous plants such as guar, soybean, lucerne, velvet bean, cowpea, moth bean, etc. The adult *O. phaseoli* is a small fly (females 2.2 mm and males 1.9 mm in length), shiny black in color except for legs, antennae and wing veins, which are light brown. Maggots are apodous and yellowish in color. Adults make punctures on young leaves and deposit eggs. On hatching maggots bore into nearest vein, reach the stem through petiole, bore down the stem, and feed on cortical layers and may extend to tap root resulting in wilting and death of plant. Young branches of grown-up plants may also be infested resulting in wilting of infested branches.

Management

As in case of *Melanagromyza sojae.*

18.2.2.3 TOBACCO CATERPILLAR: Spodoptera litura Fab. (LEPIDOPTERA: NOCTUIDAE)

It is a highly polyphagous and a major crop pest. Its host range includes many important cultivated crops such as soybean, groundnut, cotton, tobacco, castor, pulses, cabbage, cauliflower, chillies, etc.

Moths are stout, 15–20-mm long with grey-brown body and wingspan of 30–38 mm. The forewings are grey to reddish-brown with a strongly variegated pattern and paler lines along the veins; the hind wings are greyish-white with grey margins, often with dark veins. Full-grown caterpillars are stout, cylindrical, and measure 35–40 mm in length. It is velvety black with yellowish-green dorsal stripes and lateral white bands with incomplete ring-like dark band.

Female moth lays masses of eggs covered with hairs on the underside of young leaves. After hatching of eggs the caterpillars remain gregariously and feed on green tissues by scrapping from under surface of leaves. Later instars larvae are solitary and feed voraciously on leaves, stems, buds, flowers, and pods. Fecal pellets found on the leaves and on the ground indicate the presence of this pest.

Management

Collection and destruction of the infested leaves harboring egg masses and gregarious larvae helps in reducing the initial population buildup of the pests. Sunflower and castor plants may be grown as trap crops. Setting up light traps along with setting up of pheromone traps @ 12/ha for monitoring the pest. Spraying of SlNPV @ 250 LE/ha, *Bacillus thuringiensis* var *kurstaki* @ 2 g/L of water. Releases of egg parasitoid Trichogramma chilonis @ 50,000/ha/week four times reduce the population considerably. Application of neem kernel extract and *Pongamia glabra* oil during the early stages of crop growth and foliar spraying with thiodicarb 7 WP @ 2 mL/L or quinalphos 25 EC @ 2.5 mL/L or acephate 7 SP @ 1.5 g/L of water is very effective.

18.2.2.4 GIRDLE BEETLE: Obereopsis brevis Swedenbord (COLEOPTERA: CERAMBYCIDAE)

It is a major pest of soybean. Grub is yellow with dark head. The beetle makes two girdles on the stem or sometimes on petiole and lay eggs between the girdles. After hatching the grub makes a tunnel inside the stem and moves down the plant, feeding on inner contents and resulting in drying of the top portion of plant above the girdles.

Management

Deep summer plowing is required to expose the hibernating pests to natural enemies. Optimum seed rate (70–100 kg/ha) should be used. Crop rotation should be followed while avoiding excess use of nitrogenous fertilizers. Collect and destroy infested plant parts. Apply phorate 10 G @ 10 kg/ha or carbofuran 3 G @ 30 kg/ha at the time of sowing. One or two sprays of 0.03% dimethoate 30 EC or 0.05% quinalphos 25 EC can check further damage.

18.2.2.5 BIHAR HAIRY CATERPILLAR: Spilosoma obliqua Walker (LEPIDOPTERA: ARCTIIDAE)

Young larvae (first and second instars) feed gregariously on chlorophyll from the under surface of the leaves. As a result of this the leaves of the plant are skeletonized and give an appearance of net or web. Later on, third and onward instars dispersed and moved from one plant to another and eat voraciously on the older leaves, stems, shoots, flowers and pods causing serious damage to the plants.

Management

Deep summer plowing is required to expose the pupae. Intercropping soybean either with early maturing pigeon pea variety or maize or sorghum in the sequence of 4:2 should be practiced. Collection and destruction of infested plant parts along with egg masses and young larvae is done by fire. Installations of one light trap per hectare to catch the adults. Apply chlorpyriphos 20 EC @ 1.5 L/ha or trizophos 40 EC @ 0.8 L/ha or quinalphos 25 EC @ 1.5 L/ha.

18.2.2.6 GRAM CATERPILLAR: Helicoverpa armigera Hub. (LEPIDOPTERA: NOCTUIDAE)

The green larvae of *H. armigera* feed on leaves and tender shoots firstly; later on, they bore pods and feed inside (Biswas, 2013). The half portion of larvae remains inside pod while feeding on the developing seeds. They can cut hole on one to another locule and destroy 20–25 pods in its lifetime.

18.2.2.7 SOYBEAN SEMILOOPER OR GREEN SEMILOOPER: Thysanoplusia = (PLUSIA) orichalcea (F.) (NOCTUIDAE: LEPIDOPTERA)

Adults with wingspan of 36–44 mm bear a tuft of hair on thoracic region. Forewings are brown with a large, bright golden patch. The hind wings are fawn-colored, darkening toward the outer margins. Adults are active during evening. The larval body is tapering toward the head and it moves with a distinctive looping action having only three pairs of ventral prolegs. Larval color varies considerably. Larva is green with blackish head. Larvae are characterized by presence of two lateral white lines extending over the body, an additional blackish middorsal line and two other faintly marked blackish lines, subdorsal in position on either sides.

Small larvae feed on only one side of the leaf, leaving translucent "feeding windows." As larvae develop, they chew holes in the leaf and then feed from the leaf margin. Larvae are primarily foliage feeders but will attack the flowers and developing pods.

Management

Moths are attracted to light and can be trap by using light traps. Mechanical collection and destruction of larvae will be able to compensate the vigorously growing plants damage. This pest is attacked by numerous predators and parasites. Therefore, conservation of natural enemies is very helpful in keeping the pest-population under check. Early instar larvae can be controlled with *Bt* var *kurstaki* @ 0.75–1.0 kg/ha. Spraying with chlorantraniliprole 18.5% SC @ 150 mL/ha or indoxacarb 15.8% EC @ 333 mL/ha or quinnalphos 25 EC @ 1000 mL/ha is effective.

18.2.2.8 PEA POD BORER OR LIMA BEAN POD BORER: Etiella zinckenella Treitschke (LEPIDOPTERA: PYRALIDAE)

Major host plants of *Etiella zinckenella* in India include pigeon pea, cowpea, lablab, soybean, peas, chickpea, horse gram, green and black grams, *Lathyrus sativus*, etc. Adult: Body length 8–11 mm, wingspan 19–27 mm. Adult forewings are brownish-grey with a white strip along the leading edge of narrow forewings and with orange spot on basal third segments. Hind wings are transparent to opaque with darker outer edges and top of abdomen with a tuft of golden-yellow hairs. Body length of full-grown larva measures 15–22 mm. The young larvae are green, but become pinkish-red as they grow older.

Larva bores into the developing pods and feed on the seeds. Damaged pods can easily be recognized. Large pods are marked with a brown spot where the larva has entered. As the larva develops within the pod, feces accumulate causing soft, rotten patches on the pod. Seeds are either partially or entirely eaten, and considerable frass and silk can be found inside the pod.

18.2.2.9 GREEN BUG: Nezara viridula Linnaeus (HEMIPTERA: PENTATOMIDAE)

The nymphal stages are multicolored, whereas the adults are uniform green. The bugs mainly attack on fruits, and through the removal of sap and the injection of saliva, causes discoloration, malformation, stunting, and shriveling of infested fruits. In soybeans an important effect of the bugs feeding was a reduction in the germinability and oil content of seed. Attack of green bug is resulted in affected pod development, increased pod fall, and reduced number of seeds per pod.

18.2.3 PESTS OF BERSEEM (*Trifolium alexandrinum*)

Common insect pests infesting berseem are alfalfa aphid, *Therioaphis trifolii* (M.); thrips, *Thrips tabaci* (L.); whitefly, *Bemisia tabaci* (G.); cutworm, *Agrotis ipsilon* (H.); tobacco caterpillar, *Spodoptera litura* (F.); lucerne caterpillar, *Spodoptera exigua* (H.); lucerne weevil, *Hypera postica* (Gyll.); green stink bug, *Nezara viridula* (L.); gram caterpillar, *Helicoverpa armigera* (H.); bihar hairy caterpillar, *Spilosoma obliqua* (W.); hairy caterpillars, *Euproctis* spp.; green semi-loopers, *Plusia nigrisigna* (W.), *Trichoplusia ni* (H.), *Thysanoplusia orichalcea* (F.); aphid, *Aphis craccivora* (K.); dusky bug, *Oxycarenus* sp.; leaf miner, and *Biloba* (*Stomopteryx*) *subsecivella* (Zell.) are the common insect pests of berseem (Thontadarya et al., 1979; Dubey et al., 1995; Dhaliwal, et al., 1999; Saxena et al., 2002; Randhawa et al., 2009; Mari and Leghari, 2015).

18.2.3.1 POD BORER: *Helicoverpa armigera* Hub. (LEPIDOPTERA: NOCTUIDAE)

Among the pest complex of berseem *Helicoverpa armigera* (Hubner) is in the fore front (Mawar et al., 2015). It is a highly polyphagous pest and its host range includes many important cultivated crops such as cotton, sorghum, lablab, pea, chillies, groundnut, tobacco, okra, maize, tomato, soybean, safflower, gram, etc. The moth is stoutly built and yellowish brown in color. There is a dark speck and a dark area near the outer margins of each forewing. The fore wings are marked with grayish wavy lines and black spots of varying size and a black kidney-shaped mark and a round spot on the underside. The full-grown larva is stout bodied usually greenish or brownish caterpillar, measuring about 4 cm in length with longitudinal bands.

The larvae devour flower buds and developing seed of berseem. The larvae feed mostly on apical portion of the inflorescence and destroy it partially or wholly. *H. armigera* alone can reduce yield of seed crop of berseem and lucerne by 80–90% (Randhawa et al., 2009). In India 70–80% reduction in the yield of berseem seed by *H. armigera* was recorded by Balraj et al. (1975).

Management

Avoid raising of berseem seed crop adjoining to tomato, gram, late sown wheat, moong, and sunflower as because the pest multiplies on these hosts and later shift to berseem. If it is not possible, the pest should be controlled properly on these crops grown in the vicinity of berseem in order to check its migration to berseem fields. Setting up of light trap to attract and kill the moths. Set up pheromone traps @ 12 nos./ha to attract male moths and killing. Release of egg parasitoide *Trichogramma pretiosum* and egg larval parasitoide *Chelonus blackburnii*. Application of nuclear polyhedrosis virus @ 250 LE/ha. Early instar larvae can be controlled with *Bt var kurstaki* @ 0.75–1.0 kg/ha. Spraying with chlorantraniliprole 18.5% SC @ 150 mL/ha or indoxacarb 15.8% EC @ 500 mL/ha or quinnalphos 25% EC @ 1000 mL/ha or Spinosad 48% SC @ 150 mL/ha is effective.

18.2.3.2 CABBAGE SEMILOOPER: Trichoplusia ni Hubner (LEPIDOPTERA: NOCTUIDAE)

Larvae of *T. ni* were recorded causing damage to at least 160 species, varieties and cultivars in 36 families of plants. Adult *T. ni* are mottled brownish in color. The forewings with a span of about 3.8 cm, each bears an eight-shaped silvery mark near the middle. The larva has three pairs of thoracic legs and three pairs of fleshy abdominal prolegs. It crawls by forming a loop and projecting the body forward. Larvae are green with a white lateral line and two whitish lines along the middle of the dorsal surface. Pupation takes place in a loosely spun cocoon either on the underside of a leaf or in plant debris at the soil surface.

It is a polyphagous pest and its larvae cause severe damage to berseem. Larvae make round holes in the leaves and defoliate the plants.

Management

During March–April, harvest the berseem crop at regular interval (30 days to avoid lodging which creates favorable conditions for pest survival and multiplication and also hinders the activity of predatory birds which play key role to control pest. Chemical control as in case of *H. armigera*.

18.2.3.3 BIHAR HAIRY CATERPILLAR: Spilosoma obliqua Walker (LEPIDOPTERA: ARCTIIDAE)

It is a polyphagous and sporadic pest. It attacks berseem crop starting from the very beginning of the crop growth stage and devour the young crop to the ground level. In grown-up plants, also it feeds voraciously on foliage of plants and causes severe losses.

18.2.4 PESTS OF GUAR OR CLUSTERBEAN (Cymopsis tetragonaloba Linn.)

Sucking pests are the major limiting factors in fodder guar cultivation. According to Yadav et al. (2015) the sucking insect pests, namely, leafhopper (jassid), *Empoasca motti* Pruthi; whitefly, *Bemisia tabaci* (Genn.); and aphid, *Aphis craccivora* Koch inflict devastating damage in cluster bean.

The leafhopper or jassid, *E. motti* Pruthi is a serious polyphagous pest which adversely affects the vegetative growth and seed yield up to 20% (Singh, 1997). It is 3 mm long, delicate and yellowish green insect. Both nymph and adults of jassid, whitefly and aphid suck cell sap from the plants and cause heavy losses. Butani (1980) mentioned that nymphs and adults of flower thrips, *Megalurothrips distalis* (Thysanoptera: Thripidae) infest the flowers and feed on pedicles, sepals, petals, and even the stigma of the flowers of cluster bean. In cases of severe infestation, which is rare, the flowers are devitalized and shed prematurely. According to Kooner et al. (2007) triazophos 40 EC at 1.5 L/ha, ethion 50 EC at 2.0 L/ha and dimethoate 30 EC were found effective in reducing the incidence of bean thrips and they significantly increased the yield.

Dimethoate (0.03%), imidacloprid (0.005 %), and thiamethoxam (0.025 %) are the most effective insecticides against leaf hopper, whitefly, and aphid on cluster bean (Yadav et al., 2015b). According to Vadja and Kalasariya (2015) imidacloprid (0.006 %), acephate (0.15 %), clothianidin (0.025 %), difenthiuron (0.07 %), buprofezin (0.05 %), and thiacloprid (0.008 %) are effective against aphid, *Aphis craccivora* (Koch).

Mawar et al. (2015) reported Yellow mite, *Polyphagotarsonemus latus* as a pest of Guar in India.

18.2.4.1 YELLOW MITE: *Polyphagotarsonemus latus* Banks (ARACHNIDA: ACARI: TARSONEMIDAE)

The adults measure 0.1 mm in length and bear four pairs of legs. They are translucent and yellowish green in color.

The small mites live in colonies and infest mostly the top leaves. Both nymphs and adults suck sap and devitalize the plant. They are found in large numbers on the undersurface of leaves and by sucking the sap of the plant cause downward curling of the leaves.

Management

Foliar spraying with dicofol 18.5 EC @ 2 L or diafenthiuron 50 WP @ 600 g or fenazaquin 10 EC @ 1.25 L or fenpyroximate 5 EC @ 300–600 mL with 500–750 L water/ha.

18.2.5 PESTS OF METHA OR FENUGREEK (*Trigonella foenum-graecum* L.)

Stem fly: *Ophiomyia* spp. (Diptera: Agromyzidae), cowpea aphid: *Aphis craccivora* Koch (Hemiptera: Aphididae), Serpentine leaf miner: *Liriomyza trifolii* Burgess (Diptera: Agromyzidae), thrips: *Scirtothrips dorsalis* Hood (Thysanoptera: Thripidae), lucerne weevil: *Hypera postica* Gyllenhal (Coleoptera: Curculionidae), spotted pod borer: *Maruca testulalis* Geyer (Lepidoptera: Crambidae) are the common insect pests that attack fenugreek in India. Fenugreek has earlier been reported to be attacked by different species of aphids, namely, *Aphis craccivara* (Brar and Kanwar, 1994) and *Acyrthosiphon pisum* (Dadhich et al., 1989). *Aphis craccivora* is reported to cause seed yield losses to the extent of 60–68% (Sharma and Kalra, 2002). The aphid, *Acyrthosiphon pisum* population and seed yield of fenugreek had a significant inverse correlation (r = −0.94) (Naga and Kumawat, 2015). According to Shekhawat et al. (2016) minimum incidence of aphid on fenugreek was recorded with spray of NSKE @ 5.0%.

18.2.6 PESTS OF VELVET BEAN (*Mucuna pruriens*)

18.2.6.1 LABLAB BUG: *Megacopta cribraria* (FABRICIUS) (HEMIPTERA: PLATASPIDIDAE)

This pest is mainly associated with leguminous plant and common fodder hosts recorded are soybean (Thippeswamy and Rajagopal, 2005), cluster bean (Ramakrishna Ayyar, 1913), velvet bean (Rani and Sridhar, 2004), and Agathi (Srinivasaperumal et al., 1992).

Adults are 3.5–6.0-mm long, light brown to olive green with dark punctation. Fifth-instar nymphs 4–5-mm long, oval, light to dark brown, hirsute; lateral margins of thorax, and abdomen somewhat flattened.

Thippeswamy and Rajagopal (2005) reported that *M. cribraria* feeds on leaves, stems, flowers, and pods, but prefers tender new growth to older growth. They also noted that white "patches" developed at the site of feeding and later turned brownish, gradually coalescing into a necrotic area and that shoots withered with heavy infestations and bean pods did not develop normally.

18.2.6.2 SPIDER MITE: *Tetranychus ludeni* Zacher (ARACHNIDA: ACARI: TETRANYCHIDAE)

The spider mite species, *Tetranychus ludeni* is a serious pest of a wide variety of economically important plants. Larva possesses three pairs of legs and greenish yellow in color. Protonymph or the first stage nymph is an active instar characterized by the presence of four pairs of legs. It is larger in size, pale yellow in color with greenish spots on its dorsolateral region. The deutonymph or the second stage nymph is slightly larger in size than the protonymph. The adult male is much smaller in size than female, pale yellow in color and spindle shaped. Males are faster than females and are found in less number compared to female. The adult female is much larger than the male, reddish in color with cylindrically shaped abdomen (Kaimal and Ramani, 2011).

The mites often infest the upper surface of the leaves causes yellowing of leaves followed by formation of necrotic patches and drying up. It is one of the important mite pests of vegetable crops in India reported to attack French bean, brinjal, potato, water melon and many other vegetable and fruit crops limiting the production of these crops (Jeppson et al., 1975;

Puttaswamy and Channabasavanna, 1980). As a highly polyphagous mite, *T. ludeni* occurs in the field almost throughout the year.

Management

Foliar spraying with dicofol 18.5 EC @ 2 L or diafenthiuron 50 WP @ 600 g or fenazaquin 10 EC @ 1.25 L or fenpyroximate 5 EC @ 300–600 mL with 500–750 L water/ha.

18.2.7 PESTS OF MOTH BEAN (Vigna aconitifolia (Jacq.)

According to Kumar (2002) more than 20 insect-pests affect this crop from sowing to harvesting and even during storage. Puttaswamy et al. (1977) recorded 28 species of insects on moth bean in Karnataka, India. White grub, *Holotrichia* spp. (Coleoptera: Scarabaeidae); aphids, *Aphis craccivora* Koch (Hemiptera: Aphididae); spotted pod borer, *Maruca vitrata* Geyer (Lepidoptera: Crambidae); leafhopper, *Empoasca kerri* Pruthi (Hemiptera: Cicadellidae); whitefly, *Bemisia tabaci* (Gennadius) (Hemiptera: Aleyrodidae); pod bugs, *Riptortus pedestris* Fabricius (Hemiptera: Alydidae); *Clavigralla gibbosa* Spinola (Hemiptera: Coreidae), and *Nezara viridula* Linnaeus (Hemiptera: Pentatomidae) are the common insect pests of moth bean in India.

Brief account of the following moth bean pests has been provided by Kumar (2002).

18.2.7.1 JASSIDS OR LEAF HOPPER: Empoasca kerri Pruthi (HEMIPTERA: CICADELLIDAE)

This pest remains active from vegetative stage to the crop harvest. The adults as well as the nymphs, suck the cell sap. The adult is a small insect and feeds on the leaves. There are many generations of jassids during the year. In case of heavy infestation, the leaves turn brown, curl, and finally dry out and shed on the ground.

Management

Early sowing up to July 10 has resulted in good control of jassids. Intercropping of pearl millet with moth bean (1:4) has been found effective in lowering the population of jassids compared to the sole crop. Jassids can be effectively controlled by spraying monocrotophos/dimethoate (0.03%).

18.2.7.2 WHITEFLY: Bemisia tabaci (Gennadius) (HEMIPTERA: ALEYRODIDAE)

Whitefly is a serious pest of moth bean and acts as a vector for yellow mosaic virus. Incidence of whitefly is generally at peak during second week of September. The nymphs and the adults suck the cell sap particularly, from the surface of the leaves.

Management

As in case of jassids.

18.2.7.3 WHITE GRUB: Holotrichia spp. (COLEOPTERA: SCARABAEIDAE)

White grub is a serious pest damaging most of the rain fed crops including moth bean. Grubs are "C"-shaped, whitish-yellow in color with a brown head and remain close to the base of the clump. Adults are dark brown and emerge out from the soils following rains. They feed on a variety of host foliage trees. The beetles lay eggs in soil. Grubs have only one generation in one year. The grub feed on the roots from July to October. Plants show varying degree of yellowing, get wilted and their sudden death may occur.

Management

Deep plowing after summer shower and at the time of land preparation for exposing the pupae and beetles under hot sun or predatory birds. Mass collection and destruction of beetles from the branches of host plants of the beetle such as neem, subabul, *Acacia*, ber, etc. Flooding the field for 24 h for killing the grubs. Setup light traps to attract and kill the adults. Utilization of fungal pathogens such as *Metarrhizium anisopliae* and *Beauveria bassiana* in soil for causing pathogenicity to the grubs. Application of phorate 10 G @ 15 kg/ha at sowing time can effectively control the pest.

18.3 LEGUMINOUS PERENNIAL FORAGES

18.3.1 PESTS OF LUCERNE OR ALFALFA (Medicago sativa L.)

Lucerne is unique among field crops in that it is perennial and exists almost continuously over vast geographical areas and different climatic zones (Davis et al., 1974). Pea aphid (*Acyrthosiphon pisum* Harris), blue alfalfa aphid (*Acyrthosiphon kondoi* Shinjii), spotted alfalfa aphid (*Therioaphis trifolli* F.), cowpea aphid (*Aphis craccivora* Koch.), jassids (*Empoasca* spp), gram pod borer (*Helicoverpa armigera* Hub.), leaf eating caterpillar (*Spodoptera litura* Fab.), alfalfa weevil (*Hypera postica* Gyll.), Bihar hairy caterpillar (*Spilosoma obliqua* Walker), groundnut leaf miner (*Aproaerema modicella* Devanter), lucerne seed chalcid (*Bruchophagous roddi* Guss) are the common pests, sometimes causing heavy damage to Lucerne. The quantitative losses recorded in India are about 37.7% due to insect pests in lucerne (Shri Ram and Gupta, 1989).

18.3.1.1 APHIDS

Lucerne suffers damage both qualitatively and quantitatively by aphids (*Acyrthiosiphon pisum* Harris, *Acyrthiosiphon kondoi* Shinjii and *Therioaphis trifolii f. maculata* (Martin and Leonard, 1976; Golage, et al., 2011). Spotted aphid, pea aphid, and cowpea aphid are found major pest on lucerne (Anonymous, 2004). According to Mawar et al. (2015) *Acyrthosiphon pisum* and *Theriophis trifolii f.maculata* cause damage to lucerne.

18.3.1.1.1 Yellow Clover Aphid: Therioaphis trifolii Monell (Hemiptera: Aphididae)

Adults are pale yellow green in color, winged or wingless and about 1.4–2.2-mm long. They have 4–6 rows of visible tiny black spots running lengthwise on their backs. Most adults are wingless, but those with wings have smoky areas along the veins. Nymphs are similar to adults but are smaller in size. Aphids can reproduce both asexually and sexually.

Damage

Adults and nymphs suck sap from undersides of leaves and inject a toxin into the plant causing yellowing or whitening of the leaf veins, wilting of plants, and subsequently shedding of leaves. Aphids secrete honeydew which causes development of black sooty mould that inhibits photosynthesis and can decrease plant growth and palatability to stock. In severe infestation, only stems remain standing and the entire plant may die. Spotted alfalfa aphids also cause indirect damage by spreading plant viruses such as alfalfa mosaic virus in medic and lucerne pastures, and bean yellow mosaic virus in clover.

Management

Use of yellow sticky traps for attracting alate adults is required. According to Sandhu and Nijjar (1980) resistant variety and clone of lucerne were T9 and T9-85 with antibiosis mechanism of resistance and susceptible variety was LLI against *Therioaphis trifolii*. Spraying with thiamethoxam (0.005 %) and *Verticillium lecanii* 4×10^5 cfu mL^{-1} (Golage, et al., 2011).

18.3.1.1.2 PeaAphids:AcyrthosiphonpisumHarris(Hemiptera: Aphididae)

The pea aphid is a worldwide pest of economically important legume crops. These are green, yellow or pink in color with long cornicles on the abdomen. The aphids are about 0.5-cm long. These may be winged or wingless and usually only females are present. This aphid, like the other aphids feeds by sucking sap from the tender apical plant parts. The aphids excrete a lot of honeydew leading to dryness of alfalfa bush because of disruption of photosynthesis and the plants become stunted (Harris, 2006).

Management

Foliar application of single and binary combinations of vermiwash with biopesticides, minimize the infestation of the *A. pisum* and improve the crop productivity. Significant decrease in *A. pisum* population was observed after foliar spray of vermiwash with neem oil followed by aqueous garlic and annona leaf extract. The combination of neem oil with vermiwash caused complete removal of the *A. pisum* population. Vermiwash obtained from municipal solid wastes and animal dung with neem oil was found to be the most effective management against *A. pisum*. The use of vermiwash of buffalo dung and municipal solid wastes with neem oil or garlic extract is better alternative to manage the pea aphid infestation by *A. pisum* in pea crop (Mishra et al., 2015). Garlic bulbs (*Allium sativum*), Endod (*Phytolacca dodecandra*), and neem seeds (*Azadirachta indica*) extracts at 5% and 10% dilutions were found effective (Megersa, 2016).

18.3.1.2 BIHAR HAIRY CATERPILLAR: Spilosoma obliqua Walker (LEPIDOPTERA: ARCTIIDAE)

It is a highly polyphagous pest and major hosts include jute, groundnut, lucerne, sunflower, cashew, castor, cucurbits, millets, etc. It has been reported to feed on 96 plant species in India.

The full-grown larva is darkened with yellowish brown abdomen having numerous pale white, brown, and black hairs and measures about 43 mm in length. The adult is dull yellow with oblique line of black dots on forewings with a 40–50-mm wing span. The dorsal side of abdomen is red with dull yellow ventral side.

The early instar larvae remain gregarious and feed from underside of leaves by scrapping the green tissues. Grown-up larvae get dispersed, move from one field to other and vigorously feed on leaves and defoliate the plants.

Management

As in case of soybean.

18.3.1.3 LUCERNE WEEVIL: Hypera postica Gyllenhal (COLEOPTERA: CURCULIONIDAE)

The adult weevils are brown-colored stout bodied beetles measuring about 4–5.5 mm in length. A dark strip extends downwards more than half-length of the body. Full-grown last (fourth) instar grubs are about 8-mm long and light green in color with median white line along dorsal side of the body.

The injury is confined largely to the growing points, giving them a frazzled appearance showing a skeletonizing or shredding of the tips of the new growth and resulting in stunted growth of plants. The leaves thus skeletonized make poor-quality hay. Larval feeding causes damaged leaves to dry rapidly presenting a bleached out appearance, giving damaged fields the appearance of severe frost injury. Larvae cause the most significant damage particularly to seed crops of alfalfa.

Management

A total of 43 genotypes were evaluated for reaction to *H. postica* in the field. Oviposition and larval feeding were positively correlated with stem girth and plant height and negatively correlated with crown width. Genotypes exhibiting maximum resistance are B-15, B-22, B-135, B-209, Sirsa-9, Sarnac, and Dupuits (Panday et al., 1990).

The optimum time for harvesting of alfalfa crop is 50–55 days after sprouting. The damage by the pest coincides with blooming stage of the alfalfa crop. Timing of harvest is the most important manipulation to reduce weevil problems.

Spraying with neem leaf extract or chloropyriphos 20 EC @ 0.02% or dimethoate 30 EC @ 0.03% or malathion 50 EC @ 0.05% or quinalphos 25 EC @ 0.04% is effective against the pest.

18.3.1.4 GROUNDNUT LEAF MINER: Aproaerema modicella Devanter (LEPIDOPTERA: GELECHIIDAE)

The groundnut leaf miner is a widespread and frequently serious pest of groundnut, lucerne and soybean throughout south and southeast Asia. Adult is dark brown with a white spot on the coastal margin of each forewing. The small hind wings are covered by fringe of minute hair. Full-grown caterpillar is greenish or pale brown with a small dark head.

The freshly emerged larva mines the leaves along midrib causing blister-like blotches. Older larva webs the leaves together and feeds on them from within. As a result, the leaflets turn brown, shrivel, and dry up. Severely infested crop gives a burntup appearance and yield losses can reach up to 76% (Anon, 1986).

Management

Avoid water stress in irrigated crop to avoid the pest infestation. Maintain the fields and bunds free from weeds. Setup light traps between 8 and 11 p.m. at ground level. Collection and destruction of infested plant parts along with the larvae. Crop rotation with a non-leguminous crop to avoid outbreaks of the pest. Foliar sprays with acephate 1 g/L or chlorpyriphos 2.5 mL/L or monocrotophos 1.6 mL/L of water.

18.3.1.5 LUCERNE SEED CHALCID: Bruchophagous roddi Guss (HYMENOPTERA: EURYTOMIDAE)

B. roddi only attacks the seeds of Medicago plants (Zerova, 1995) and it can be found where alfalfa is grown (Grigorov, 1976). The adult alfalfa seed chalcid is a minute, black wasp. Only some parts of the legs (on the tibia and tarsus) are yellow-brown. The male is 1.2–1.7 mm and the female is 1.3–1.8-mm long. The fully grown larva (grub) is white, apodous, "C"-shaped, and approximately equal to the size of an alfalfa seed (1.5–2 mm) (Grigorov, 1976 and Padmavathi et al., 2003).

Female wasp inserts its eggs singly into the developing seed through green pod. The developing grub feeds on the internal contents of the seed leaving only the seed coat. The grub develops within the seed and pupates there itself. Emergence hole of the adult is visible on seed coat and the pod.

Management

Clipping back established stands helps to delay bloom, provides a shorter pollination window and reduces the time that the green pods are available for egg laying. Two treatments with 0.1% fenitrothion or 0.2% carbaryl or 2% diazinon one applied at budding and the other at the green-pod stage, reduce the percentage of infested seeds.

18.3.2 PEST OF STYLOS (Stylosanthes spp.)

Feeding of blister beetles, *Mylabris pustulata*, *Mylabris thunbergii*, and another *Mylabris* sp. on *S. hamata* were recorded from Tamil Nadu and Uttar Pradesh in India (Durairaj and Ganapathy, 2003). Flowers are eaten away irregularly by the adults. Pod borer (*Helicoverpa armigera*), hairy caterpillars (*Spilosoma obliqua*), leaf hopper (*Empoasca kerii*), and black aphid (*Aphis craccivora*) also cause damage to this crop. Severe incidence of *Helicoverpa armigera* on *S. scabra* was recorded. The larvae were found feeding on the spike, floral parts, and young ovules (Ramesh et al., 1997).

18.4 LEGUMINOUS PERENNIAL TREES

18.4.1 PESTS OF AGATHI (*Sesbania grandiflora* L.)

Megacopta cribraria (Fabricius) (Hemiptera: Plataspididae) is a serious sucking pest on *Sesbania grandiflora* (Srinivasaperumal et al., 1992). Another sap sucker, *Ceroplastodes* sp. (Hemiptera: Coccidae) has been reported from Andaman and Nicobar Islands (Shah et al., 1989). The leaf eating caterpillar complex on *Sesbania* included *Eurema hecabae* (Wallace) (Lepidoptera: Pieridae), *Hyposidra successaria* (Walker) (Lepidoptera: Geometridae), *Maruca testulalis* (Geyer) (Lepidoptera: Crambidae), *Spodoptera litura* (Fab.) (Lepidoptera: Noctuidae), and *Helicoverpa armigera* (Hub.) (Lepidoptera: Noctuidae). *E. hecabae* is the major pest on *sesbania* during the first few months of the crop growth (Sireesha et al., 2009). Nayar et al. (1976) also reported leaf eating caterpillars on *sesbania* which included *Semiothisa pervolgata* (Walker) (Lepidoptera: Geometridae) and *S. litura* (Noctuidae: Lepidoptera). Incidence of stem borer, *Azygophleps scalaris* Fabr has also been recorded on *Sesbania* (Reghupathy et al., 1997, and Sireesha et al., 2009). Moderate level of infestation on Sesbania by papaya mealybug, *Paracoccus marginatus* (Williams and Granara de Willink) (Hemiptera: Pseudococcidae) was reported by Selvaraju and Sakthivel (2011). Ananthakrishnan (1971) reported that heavy infestations of *Megalurothrips distalis* (Thysanoptera: Thripidae) in the flowers of *Sesbania grandiflora* with over 150–200 individuals per flower caused their complete drying, withering, and shedding.

18.4.2 PESTS OF SUBABUL (*Leucaena leucocephala* Lam.)

18.4.2.1 STRIPED MEALYBUG: *Ferrisia virgata* Cockerell (HEMIPTERA: PSEUDOCOCCIDAE)

F. virgata is one of the most highly polyphagous mealybugs known attacking plant species belonging to some 203 genera in 77 families (García et al., 2016). Many of the host species belong to the Leguminosae and

Euphorbiaceae. Among the hosts of economic importance are avocado, banana, betel vine, black pepper, cashew, citrus, cocoa, coffee, custard apple, grapevine, guava, *Leucaena,* and tomato.

Adult females are oval, up to 5-mm long, grayish-yellow, with two longitudinal, submedian, interrupted dark stripes on the dorsum showing through the waxy secretion; hence the common name "striped mealybug." The dorsum also bears numerous straight, glassy threads of wax up to 4.0–4.5-mm long. Both adult females and nymphs suck sap from stems, branches, leaves, and fruits causing wilting and defoliation

Management

Severely infested branches should be cut and burnt immediately. Release of predatory coccinellid *Cryptolaemus montrouzieri.* Spraying with Diazinon, malathion, and dimethoate are effective.

18.4.2.2 SUBABUL PSYLLID: Heteropsylla cubana Crawford (HEMIPTERA: PSYLLIDAE)

It is a major pest of *Leucaena leucocephala.* It is native to Central and South America. It invaded subabul plantations in India in 1988 (Gopalan et al., 1988). The adult psyllid is difficult to see with the naked eye and is about 2-mm long and yellow in color. Nymphs are similar to adults in appearance except they are smaller and wingless (Shivankar and Rao, 2010). Eggs are laid between new leaves on young shoot tips. The insect is common on the young growth of *Leucaena* trees where the eggs, wingless nymphs, and winged adults often occur together (Senthilkumar and Murugesan, 2015). Host plants other than subabul are *Leucaena trichodes*, *Leucaena pulverulenta*, *Leucaena diversifolia*, *Leucaena salvadorensis*, and *Samanea saman* (Nair, 2001).

Both adults and nymphs suck sap from the terminal leaves, buds, and flowers of host plants. Repeated attacks cause wilting, defoliation, branch dieback, or death of host trees.

Management

Measures aimed at controlling the *Leucaena* psyllid have primarily concentrated on the development of resistant *Leucaena* varieties and the use of biological control agents. Biological control agents for the *Leucaena* psyllid include the predators, *Curinus coeruleus* and *Olla v-nigrum* (Coleoptera: Coccinellidae), and the parasitoids, *Psyllaephagus yaseeni* (Hymenoptera: Encyrtidae) and *Tamarixia leucaenae* (Hymenoptera: Eupelmidae) (Shivankar and Rao, 2010).

18.4.3 PESTS OF GLIRICIDIA (Gliricidia sepium (Jacq.) Steud.)

18.4.3.1 TEAK SAPLING BORER: Sahyadrassus malabaricus Moore (LEPIDOPTERA: HEPIALIDAE)

It is a borer pest of *Gliricidia sepium* in India (Devasahayam et al., 1987). Existing information on the life history and habits of this insect is available from published works of Beeson (1941), Nair (1982), and Nair (1987).

S. malabaricus has a wide host range of over 40 species of woody shrubs and trees belonging to 22 families, of which Ulmaceae, Verbenaceae, Mimosaceae, and Myrtaceae contain the most commonly attacked species.

Full-grown larvae measure 6–10 cm in length. They are yellowish-white in color with deep black head capsule. The first thoracic segment, parts of the 2nd and 3rd thoracic segments and some dorsal sclerites on the abdomen are brownish. The moths are large and grayish-brown in color, with characteristically mottled forewings. Female moth is about 5.5-cm long with a wingspan of 11 cm.

The larvae cause damage to saplings of various tree species by boring into the stem, often leading to breaking off of stem at the point of attack. Attacked saplings can be easily recognized by the dome-shaped mass of woody particles covering the point of attack. On removal of this cover, a large borer hole can be seen which extend down along the central core of the stem. The larva feeds only on callus growth in the vicinity of the tunnel mouth. In rare instances, the stem is ring-barked, resulting in drying up of the sapling or the stem breaks off at the point weakened by feeding (Nair, 1982).

Management

Generally, it is difficult to control borers because insecticides cannot reach their concealed habitat easily. Methods recommended against this borer in the past included physical killing with a wire probe, injection of insecticide into the tunnel, and tar plugging. Cultural practices may include avoiding excessive weed cover, particularly during June, July, and August keeping trap crops such as *Trema orientalis* and *Clerodendrum viscosum.* Application of insecticide such as quinalphos at the tunnel mouth after pulling off the particle mat cover. *Beauveria bassiana* is an effective pathogen of *S. malabaricus*.

18.4.4 PESTS OF Prosopis cineraria (L.) Druce AND P. juliflora (Sw.) DC.

As many as 154 insect species belonging to 113 genera and 48 families under eight insect orders, namely, Coleoptera, Hemiptera, Diptera, Hymenoptera, Isoptera, Lepidoptera, Orthoptera, Thysanoptera, and one Acarine pest species have been recorded feeding on *Prosopis cineraria* (69 species) and *P. juliflora* (28 species) and other *Prosopis* species (68 species) from all over the world. Out of these, 96 insect species belonging to eight orders have been recorded from Indian subcontinent. Apart from insect fauna, only one species of noninsect pest, *Eriophyes prosopidis* (Eriophyidae: Acarina) has also been recorded to cause severe damage by inducing leaf and inflorescence galls in *P. cineraria* in Rajasthan. Majority of insects are polyphagous and only causally infest *P. cineraria* and *P.juliflora* but some are potential pests and occasionally cause epidemics in the plantations (Beeson, 1941; Mathur and Singh, 1960; Singh and Bhandari, 1986; Parihar, 1993; Parihar and Singh 1993; Sharma, 2016).

Twentysix potential pests on *P. cineraria* and *P. juliflora* have been recorded from arid and semi-arid areas of Rajasthan. Of them, four species of long horned beetles, namely, *Derolus iranensis* = (*descicollis*) Pic, *Aeolesthes holosericea* Fab, *Hypoeschrus indicus Gahan*, and *Acanthophorus serraticornis* (Oliver) have been recorded to exhibit severe damage to *Prosopis cineraria* in four northwestern districts of Rajasthan (Ahmed et al., 2004).

The white grub, *Holotrichia consanguinea* (Blanchard) (Coleoptera: Scarabaeidae) causes severe injury to the seedlings. The grubs cause large scale damage in the nursery stage and the adult beetles feed on the foliage in arid areas of India (Sharma, 2016).

The larvae of *Taragama* = (*Streblote*) *siva* Lefèbvre (Lepidoptera: Lasiocampidae), are found to be the most notorious pest of *Prosopis juliflora* and cause complete defoliation of the plants in case of epidemic areas (Sharma, 2016).

Both nymphs and adults of the Desert Locust, *Schistocerca gregaria* Forskal (Orthoptera: *Acrididae*) feed on the foliage and cause moderate infestation to *Prosopis* seedlings in nurseries and young plantations (Sharma, 2016). The Aak grasshopper, *Poekilocerus pictus* Fabricius (Orthoptera: Pyrgomorphidae) is a primary pest of Aak (*Calotropis* sp.) but its infestation has also been reported on *Prosopis cineraria*. The nymphs and adults were seen feeding on the foliage, making irregular holes on *P. cineraria* (Kumar et al., 1994). Recently *P. pictus* has been recorded to attack severely large number of seedlings of *P. cineraria* and *P. juliflora* in various nurseries (Sharma, 2016).

David and Subramaniam (1976) reported *Acaudaleyrodes rachipora* Singh (Hemiptera: Aleyrodidae), to cause injury to *Prosopis spp.*, from South India. This species is highly polyphagous in nature and infestation in the nurseries results in appearance of chlorotic spots at feeding sites on leaf surfaces. Vast amount of honeydew is produced by the pest which leads to development of sooty mould on leaves, and adversely affects photosynthesis leading to growth retardation and reduced vigour. (Sundararaj and Murugesan, 1996).

Parihar (1993), Kumar et al. (1994), Yousuf, and Gaur (1998) have reported *Oxyrachis tarandus* Fab. (Hemiptera: Membracidae), *O. rufescens*, and *Eurybrachys* sp. (Hemiptera: Eurybrachidae), infesting *P. cineraria*, and *P. juliflora* from arid areas throughout the active monsoon. *Oxyrachis tarandus* lay eggs on shoots in a V-shaped slit and injury often results in the stunting or ultimate death of the infested shoot. The nymphs and adults feed gregariously on the sap of the tender shoots. The insects excrete honeydew which is usually attended by ants (Sharma, 2016).

Eurybrachys tomentosa Fabricius was recorded infesting *P. cineraria* in arid and semi-arid areas. Newly hatched nymphs are gregarious in nature. In case of severe infestation of this leafhopper the growth of new shoots is checked, leaves are shed and young shoot died. *Nezara viridula* Linnaeus (Hemiptera: Pentatomidae), a polyphagous sap-sucking bug, has also been recorded to infest seedlings of *P. juliflora* and *P. cineraria* in arid areas (Sharma, 2016).

Homoeocerus variabilis Dallas (Hemiptera: Coreidae) is one of the common pest that damage the tree rapidly. The incidence of growing rate of the bug is highest in December and minimum in July. The bugs sucked the sap from newly emerging leaves, young branches, and flowers, which led to the suppression of growth of the khejri tree (*Prosopis cineraria*) through drying of branches, leaves, and flowers. The affected trees produced defective pods unsuitable for human consumption (Haldhar, 2012).

The nymphs and adults of *Frankliniella schultzei* Trybom (Thysanoptera: Thripidae) were reported to infest *P. cineraria and P. juliflora* in arid areas (Parihar, 1993; Murugesan and Kumar, 1996). All parts of the flowers are infested and several adults and nymphs were found inside each flower. Heavy infestations result in drying and wrinkling of inflorescence.

Three species of termites were observed to infest *Prosopis* spp. They cause a considerable damage to the dead wood as well as living trees. *Odontotermes obesus* and *Microtermes obesi* (Isoptera: Termitidae) are recorded to be the most injurious in nurseries and plantations of *Prosopis* spp. (Parihar, 1993 and Kumar et al., 1994). The infestation caused by *Microtermes mycophagous* (Isoptera: Termitidae), has also been reported by Parihar and Singh (1998) in the young plantations of *Prosopis* spp. They forage from their nest by means of runways to dead wood or to the roots and bark of living plants.

KEYWORDS

- **pests**
- **forage crops**
- **legumes**
- **symptoms**
- **management**

REFERENCES

Ahirwar, R.; Devi, P.; Gupta, R. Seasonal Incidence of Major Insect-pests and Their Biocontrol Agents of Soybean Crop (*Glycine max* L. Merrill). *Sci. Res. Essays.* **2015,** *10* (12), 402–406.

Ahmed, S. I.; Chaudhuri, K. K.; Sharma, M.; Kumar, S. New Insect Pest Records of Khejri and Rohida from Rajasthan and Their Possible Management Strategies. *Indian Forester* **2004,** *130* (12), 1361–1374

Ananthakrishnan, T. N. Thrips (Thysanoptera) in Agriculture, Horticulture and Forestry-diagnosis, Bionomics and Control. J. Sci. Ind. Res. **1971,** 30 (3), 113–46.

Anonymous. Proceedings of VII Annual Rabi-Summer Groundnut Workshop, Mahatma Phule Agriculture University Rahuri, Ahmednagar. 1986, pp 4–5.

Anonymous. 2004 Annual Report Part-2 Rabi 2001-02. AICRP on Forage Crops, IGFRI, Jhansi, September, 2004. pp 47–65.

Ashigar, M. A.; Umar K. M. Biology of *Maruca vitrata* (Lepidoptera: Crambidae), a Serous Pest of Cowpea and Other Legume Crops: A Review. *Ann. Exp. Biol.* **2016,** *4* (2), 33–37.

Balraj, S.; Chahal, B. S.; Gurmeet, S. Chemical Control of *Helicoverpa armigera* (Hbm.) (Lepidoptera; Noctuidae) on Berseem, *Trifolium alexandrium* L. *Indian J. Entomol.* **1975,** *35* (4), 285–288.

Beeson, C. F. C. The Ecology and Control of Forest Insects of India and the Neighbouring Countries, 1961 Reprint, Government of India.1941; p 767.

Biswas, G. C. Insect Pests of Soybean (*Glycine max* L.), Their Nature of Damage and Succession with the Crop Stages. *J. Asiat. Soc. Bangladesh Sci.* **2013,** *39* (1), 1–8.

Brar, K. S.; Kanwar, J. S. Field Response of Fenugreek Germplasm to *Aphis craccivora* (Koch.). *J. Insect Sci.* **1994,** *7* (2), 211–212.

Butani, D. K. Insect Pests of Vegetables and Their Control: Cluster Beans. Pesticides **1980,** 14 (10), 33–35.

Dadhich, S. R.; Kumawat, K. C.; Jain, P. C.; Sharma, J. K. In *Studies on Varietal Tolerance to Fenugreek (Trigonella foenumgraecum* L) *to Aphids and Its Management Through Insecticides.* First National Seminar on Seed Spices. 1989, pp 43–44.

David, B. V.; Subramaniam, T. R. Studies on Some Indian Aleyrodidae. *Rec. Zool. Surv. India*, **1976,** *70*, 133–233.

Davis, D. W.; Nichols, M. P.; Armbrust, E. G. The Literature of Arthropods Associated with Alfalfa. 1. A Bibliography of the Spotted Alfalfa Aphid, *Therioaphis maculata* (Buckton) (Homoptera: Aphidae). *Illinois Natural History Survey and Biological Notes No. 87.* 1974; p 14.

Devasahayam, S.; Premkumar, T.; Koya, K. M. A. Record of *Sahyadrassus malabaricus* (Moore) Damaging *Gliricidia maculata*, A Standard of Black Pepper *Piper nigrum* in Kerala. *Entomon* **1987,** *12*, 391–392.

Dhaliwal, J. S., Gupta, B. K.; Singh, A. Effect of Insect Pests Infestation on the Quality of Some Forage Crops. *Indian J. Anim. Nutr.* **1999,** *16* (2), 140–143.

Dubey, O. P.; Odak, S. C.; Gargav, V. P. Population Dynamics of Gram Pod Borer. *JNKVV Res. J.* **1995,** *27* 59–63.

Durairaj, C.; Ganapathy, N. Host Range and Host Preference of Blister Beetles. *Madras Agric. J.* **2003,** *90* (1–3), 108–114.

Gangrade, G. A. Assessment of Effect on Yield and Quality of Soybean Caused by Major Arthropod Pest, Terminals Technical Report on the Project, J.N.K.V.V., Jabalpur, MP. 1976, p 143.

Golage, G. R.; Gosavi, S. R.; Wankhede, S. M. Integrated Pest Management for Aphid in Lucerne. *Int. J. Pl. Protec.* **2011,** *4* (1), 196–198.

Gopalan, M.; Jayaraj, S.; Ariavanam, M.; Pillai, K.; Subba Rao, P. V. New Record of *Heteropsylla cubana* Crawford (Psyllidae: Homoptera) on Subabul, *Leucana leucocephala* (LAM) DE WIT in India. *Current Sci.* **1988,** *57* (20), 1124–1125.

Grigorov, St. Lucernov semeyad *Bruchophagus roddi* Guss. In *Special Entomology*, 2nd ed.; Zemizdat: Sofia, Bulgaria, 1976, pp 173–174.

Haldhar, S. M. Report of *Homoeocerus variabilis* (Hemiptera: Coreidae) on Khejri (*Prosopis cineraria*) in Rajasthan, India: Incidence and Morphometric Analysis. *Florida Entomol.* **2012,** *95* (4), 848–853.

Harris, D. Development and Testing of 'On-farm' Seed Priming. *Adv. Agronomy 90* 129–178.

Jeppson, L. R., Keifer, H. H.; Baker, E. W. *Mites Injurious to Economic Plants*. University of California Press, 1975; p 614.

Kaimal, S. G.; Ramani, N. Biology of *Tetranychus ludeni* Zacher (Acari: Tetranychidae): A Pest of Velvet Bean. *Indian J. Fundam. Appl. Life Sci.* **2011,** *1* (3) 1–6.

Kooner, B. S.; Cheema, H. K.; Taggar, G. K. Efficacy of Different Insecticides as Foliar Sprays Against Bean Thrips, M*egalurothrips distalis* (Karny) in Mungbean. *Acta. Hort.* (ISHS). **2007,** *752*, 531–534.

Kotadia, V. S.; Bhalani, P. A. Residual Toxicity of Some Insecticides Against *Aphis craccivora* Koch on Cowpea Crop. *GAU Res. J.* **1992,** *17* (2), 161–164.

Kumar, D. Production Technology for Moth Bean in India. Indian Council of Agricultural Research; Central Arid Zone Research Institute: Jodhpur, 2002; p 29.

Kumar, D.; Narain. *Production Technology for Cow pea*, ACIRP on Arid Legumes; CAZRI. Jodhpur, Rajasthan, 2005.

Kumar, S.; Ahmed, S. I.; Kumar, S. Annotated List of Insect Pests of Forest Trees of Arid and Semi Arid Regions Rajasthan and Gujarat. *Oikoassay* **1994,** *11* (1 and 2), 5–9.

Kumar, A.; Kumar, A.; Satpathy, S.; Singh, S. M.; Lal, H. Legume Pod Borer (*Maruca testulalis* Geyer) and Their Relative Yield Losses in Cowpea Cultivars. *Prog. Hortic.* **2013,** *45* (1), 229–232.

Mari, J. M.; Leghari, M. H. Biodiversity of Insect Species on Berseem Ecosystem. *Pak. J. Agri. Agril. Engg. Vet. Sci.* **2015,** *31* (1), 71–80.

Martin, J. H.; Leonard, W. H. Principles of Field Crop Production. Legume MacMillan Publication, New York. 1976; pp 577–644.

Mathur, R. N.; Singh, B. A List of Insect Pests of Forest in India and the Adjacent Countries. *India Forest Bulletin*, 1960. *171*, 1–30.

Mawar, R., Mall, A. K.; Kantwa, S. REco-friendly Plant Protection Technologies for Forages. *Bio. Evolution.* **2015**, *2* (1), 31–37.

Megersa, A. Botanicals Extracts for Control of Pea Aphid (*Acyrthosiphon pisum*; Harris). *J. Entomol. Zoo. Stud.* **2016,** *4* (1), 623–627.

Mishra, K.; Singh, K.; Tripathi, C. P. M. Combined Efficacy of Biofertilizer with Different Biopesticides on Aphid's Infestation (*Acyrnthosiphon pisum*) and the Productivity of Pea Crop (*Pisum sativum*). *Acad. J. Entomol.* **2015,** *8* (3), 110–116.

Murugesan, S; Kumar, S. New Record and Damage of Flower Thrips in the Introduced Tree Species of Arid Region. *Indian Forester* **1996,** *122* (9), 854–855.
Naga, K. C.; Kumawat, K. C. Estimation of Economic Decision Levels of Aphid, *Acyrthosiphon pisum* (Harris) on Fenugreek, *Trigonella foenum graecum* Linn. *Annal. Plant Prot. Sci.* **2015,** *23* (1), 37–42.
Naik, C. M.; Swamy, M.; Chandrappa, M.; Sasivihalli, P. B. Insect Pests of Soybean *Glycine max* (L.) Merrill. *Insect Environ.* **2013,** *19* (2), 99–100.
Nair, K. S. S. Pest Outbreaks in Tropical Forest Plantations: Is There a Greater Risk for Exotic Tree Species? Centre for International Forestry Research (CIFOR): Bogor, Indonesia, 2001, pp 74.
Nair, K. S. S. Seasonal Incidence, Host Range and Control of the Teak Sapling Borer, *Sahyadrassus malabaricus*. KFRI Research Report 16. Kerala Forest Research Institute, Peechi, Thrissur, 1982, p 36.
Nair, K. S. S. Life History, Ecology and Pest Status of the Sapling Borer, *Sahyadrassus malabaricus* (Lepidoptera, Hepialidae). Entomon **1987,** *12* (2), 167–173.
Padmavathi, C.; Pandey, K.; Rakesh, S.; Seth, R. Assessment of Losses Caused by Seed Chalcid, *Bruchophagous roddi* Guss. (Hymenoptera: Eurytomidae) in Lucerne Grown for Seed. *Indian J. Plant Prot.* **2003,** *31*,152–153.
Panday, K. C.; Singh, A.; Faruqui, S. A. Oviposition and Feeding by Lucerne weevil, *Hypera postica* Gyll. in Lucerne Genotypes. *Indian J. Genet. Plant Breed* **1990,** *50* (1), 77–80.
Pandey, S. N.; Singh, R., Sharma, V. K.; Kanwat, P. W.; Losses Due to Insect Pests in Some Kharif Pulses. *Indian J. Ent.* **1991,** *53*, 629–631.
Parihar, D. R. Insect Fauna of Khejri, *Prosopis cineraria* of Arid Zone. *Indian J. Forest* **1993,** *16*, 132–137.
Parihar, D. R.; Singh, M. P. Insects Associated with *Prosopis cineraria* in Arid Western Rajasthan, India. Proceeding of *Prosopis* Species in the Arid and Semiarid Zones of India, a Conference. Central Arid Zone Research Institute, Jodhpur, Rajasthan, India. 21–23 Nov. 1993, pp 1–115.
Parihar, D. R.; Singh, M. P. Insects Associated with *Prosopis cineraria* in arid Western Rajasthan, India. In *Prosopis specoes in the arid and semi arid zones of India*, Tewari, J. C., Pasiecznik, N. M., Harsh, L. N., Harris, P. J. C., Eds. *Prosopis* Society of India and the Henry Double Day Research Association, 1998; pp 99–102.
Patel, S. K. B.; Patel, H., Korat, D. M.; Dabhi, M. R. Seasonal Incidence of Major Insect Pests of Cowpea, *Vigna unguiculata* (Linn.) Walpers in Relation to Weather Parameters. *Karnataka J. Agric. Sci.* **2010,** *23* (3), 497–499.
Puttaswamy.; Channabasavanna, G. P.; Life history of *T. ludeni* (Acari: Tetranychidae) Under Field Condition. *Indian J. Acarol.* **1980,** *4* (1), 41–48.
Puttaswamy Gowda, B. L. V.; Ali, T. M. M. Record of Pests Infesting Moth Bean (Matki) (*Phaseolus aconitifolius* Jacq.) a Potential Pulse Crop. *Current Res.* **1977,** *6* (4), 69–71.
Raheja, A. K. Assessment of Losses Caused by Insect Pests to Cowpea in Northern Nigeria. *PANS* **1976,** *22*, 229–233.
Rai, R. K.; Patel, R. K. Girdle Beetle, *Obereopsis brevis* Swed. Incidence in *kharif* Soybean. Odisha J. Agricultural Res. **1990,** *3* (2), 163–165.
Ram, S.; Patil, B. D.; Purohit, M. L. Seasonal Incidence of Major Insect Pests of Fodder Cowpea (*Vigna unguiculata* L.) in Bundelkhand Region. *Bull. Entomol.* **1989,** *30* (1), 44–47.

Ramakrishna Ayyar, T. V. On the Life History of *Coptosoma cribraria* Fabr. *J. Bombay Nat. Hist. Soc.* **1913,** *22*, 412–414.

Ramesh, C. R.; Mal, B.; Hazra, C. R.; Sukanya, D. H.; Ramamurthy, V.; Chakraborty, S. Status of Stylosanthes Development in Other Countries. III. Stylosanthes Development and Utilization in India. *Trop. Grassl.* **1997,** *31*, 467–475.

Randhawa, H. S.; Aulakh, S. S.; Bhagat, I.; Chhina, J. S. Efficacy of Different Insecticides Against *Helicoverpa armigera* (hubner) (Lepidoptera: Noctuidae) on Seed Crop of Berseem in Punjab. *Legume Res.* **2009,** *32* (2), 145–148.

Rani, B. J.; Sridhar, V. Record of Arthropod Pests on Velvet Bean, *Mucuna pruriens* var. *utilis* Under Bangalore Conditions. *J. Med. Aromat. Plant Sci.* **2004,** *26*, 505–506.

Reghupathy, A.; Palanisamy, S.; Chandramohan, N.; Gunathilagaraj, K. A Guide on Crop Pests. Sooriya Desktop Publishers, Coimbatore, 1997; p.170.

Saini, M. K.; Singh, P.; Kular, J. S. Evaluation of New Insecticides Against *Helicoverpa armigera* (Hubner) in Berseem Seed Crop Under Punjab Conditions. *J. Insect Sci.* **2013,** *26* (2): 184–185.

Sandhu G. S. Some Insects Recorded as Pests of Fodder Crops at Ludhiana (Punjab). *J. Res. Punjab Agric. Univ.* **1977,** *14*, 449–459.

Sandhu, G. S.; Nijjar, A. S. Biology of Spotted Alfalfa Aphid, *Therioaphis trifolii* (Monell) on Resistant and Susceptible Varieties/Clone of Lucerne. *Indian J. Entomol.* **1980,** *42* (3), 398–402.

Sandhu, G. S. Note on the Incidence of *Stomopteryx subsecivella* (Zell.) on Lucerne at Ludhiana. *Indian J. Agric. Sci.* **1978,** *48*, 53–54.

Saxena, P.; Pandey, K. C.; Padmavathi, C.; Shah, N. K.; Faruqui, S. A.; Roy, S.; Hasan, N.; Bhaskar, R. B.; Azmi, M. I. Forage Plant Protection. Indian Grassland and Fodder Research Institute: Jhansi, 2002; p 38.

Selvaraju, N. G.; Sakthivel, N. Host Plants of Papaya Mealybug (*Paracoccus marginatus* Williams and Granara de Willink.) in Tamil Nadu. *Karnataka J. Agric. Sci.* **2011,** *24* (4), 567–569.

Senthilkumar, N.; Murugesan, S. Insect Pests of Important Trees Species in South India and Their Management Information. Institute of Forest Genetics and Tree Breeding (IFGTB). Indian Council of Forestry Research & Education: Forest Campus, R.S. Puram, Coimbatore, Tamil Nadu, 2015; p 132.

Shah, N. K.; Belavadi, V. V.; Pol, R. N. Occurrence of the Scale Insect *Ceroplastodes* sp. (Homoptera: Coccidae) on *Sesbania. J. Andaman Sci. Assoc.* **1989,** *5*, 86.

Shaonpius, M.; Charanjit, K. G. Influence of Weather Parameters on the Incidence of Cowpea Golden Mosaic Virus and Its Vector, *Bemisia tabaci* (Gennadius). *Indian J. Ent.* **2010,** *72* (1), 79–83.

Sharama, H. C. Bionomics, Host Plant Resistance and Management of the Legume Pod Borer, *Maruca vitrata*: A review. *Crop Prot.* **1998,** *17* (5), 373–386.

Sharma, M. Insect Pests of Forestry Plants and Their Management. *Int. J. Adv. Res.* **2016,** *4* (8), 2099–2116.

Sharma, S. S.; Kalra, V. K. Assessment of Seed Yield Losses Caused by *Aphis craccivora* Koch, in Fenugreek. *Forage Res.* **2002,** *28*, 183–184.

Shekhawat, K. S.; Shivran, A. C.; Singh, D.; Mittal, G. K.; Singh, B. Integrated Management of Diseases and Pest Through Organic Farming Approaches in Fenugreek (*Trigonella foenum-graecum* L.). *Int. J. Seed Sp.* **2016,** *6* (1), 39–42.

Shivankar V. J.; Rao. C. N. Psyllids and Their Management. *Pest Manag. Hortic. Ecosyst.* **2010,** *16* (1), 1–4.

Shri Ram.; Gupta, M. P. Integrated Pest Management in Forage Crop. National Symposium on Strategy for Forage Production and Improvement by 2000 A.D. Sept. 21–23, 1989; KAU Trivendrum, 1989.

Singh, P.; Bhandari, R. S. Insect Pests of *Prosopis* and their Control. In Proceedings of. National Symposium on the Role of *Prosopis* in Wasteland Development. Wasteland Board, Min. Environ. Forest. Government of India, New Delhi, 1986.

Singh, S. P. *Insect Pest Management in Forage Crops*, Proceeding of Advanced Training Course on Insect Pest Management, 10–29, March 1997, Department of Entomology, CCS Haryana Agricultural University, Hisar, 1997, pp 132–135.

Singh, S. R.; Allen, D. J. Pests, Diseases, Resistance and Protection in Cowpea. *Advances in Legume Science*. Summerfield, R. J., Bunting, H. H., Eds.; Ministry of Agriculture, Fisheries and Food, London: Royal Botanical Garden, Kew, 1980, pp.419–433.

Singh, S. R.; Van Emden, H. F. Insect Pests of Grain Legumes. *Ann. Rev. Ento.* **1975,** *24*, 255–278.

Sireesha, K., Ramadevi, P.; Tanuja priya, B. Biodiversity of Insect Pests and Their Natural Enemies in Betelvine Ecosystem in Andhra Pradesh. *Karnataka J. Agric. Sci.* **2009,** *22* (3), 727–728.

Srinivasaperumal, S.; Samuthiravelu, P.; Muthukrishnan, J. Host Plant Preference and Life Table of *Megacopta cribraria* (Fab.) (Hemiptera: Plataspidae). *Proc. Indian Natn. Sci. Acad.* **1992,** *B58* (6), 333–340.

Subhasree S.; Mathew M. P. Eco-friendly Management Strategies Against Pod Borer Complex of Cowpea, *Vigna unguiculata* var. *Sesquipedalis* (l.) Verdcourt. *Indian J. Fundam. Appl. Life Sci.* **2014,** *4* (4), 1–5.

Sundararaj, R.; Murugesan, S. Occurence of *Acaudaleyrodes rachipora* (Singh) (Aleyrodidae: Homoptera) as a Pest of Some Important Forest Trees in Jodhpur (India). *Indian J. Forest.* **1996,** *19* (3), 247–248.

Thippeswamy, C.; Rajagopal, B. K. Comparative Biology of *Coptosoma cribraria* Fabricius on Field Bean, Soybean and Redgram. *Karnataka J. Agric. Sci.* **2005,** *18* (1), 138–140.

Thontadarya, T. S.; Rao, K. J.; Kumar, N. G. Occurrence of the Groundnut Leaf Miner, *Biloba* (*Stomopteryx*) *subsecivella* (Zeller) on Berseem (*Trifolium alexandrinum* Linnaeus) in Karnataka. Current Res. **1979,** *8* (4), 65.

Uttam, K.; Sharma, P.; Shrivastava, S. Spectrun of Insect Pest Complex of Soybean (*Glycine max* (L) Merill) at Lambapeepal Village in Kota Region, India. *J. Bio. Sci.*, **2012,** *1*(1): 80–82.

Vadja, D. J.; Kalasariya R. L. Bio-efficacy of Newer Insecticides Against Aphid, (*Aphis craccivora* Koch) on Clusterbean. *AGRES–Int. e-J.* **2015,** *4* (2), 125–130.

Yadav, K. Swathi, Pandya, H. V.; Patel, S. M.; Patel, S. D.; Saiyad, M. M. Population Dynamics of Major Insect Pests of Cowpea [*Vigna ungiculata* (L.) Walp.]. *Int. J. Plant Prot.* **2015,** *8* (1), 112–117.

Yadav, S. R.; Kumawat, K. C.; Khinchi, S. K. Efficacy of New Insecticide Molecules and Bioagents Against Sucking Insect Pests of Cluster Bean, *Cyamopsis tetragonoloba* (Linn.) Taub. *Legume Res.* **2015b,** *38* (3), 407–410.

Yousuf, M.; Gaur, M. Some Note Worthy Insect Pests of *Prosopis juliflora* from Rajasthan. In *Prosopis Specoes in the Arid and Semi Arid Zones of India*. J.C. Tewari, N.M. Pasiecznik, L. N. Harsh, P.J.C. Harris, Eds.; *Prosopis* Society of India and the Henry Double Day Research Association: India, 1998, pp 91–94.

Zerova, M. D. The Parasitic Hymenoptera: *Subfamilies Eurytominae and Eudecatominae (Chalcidoidea, Eurytomidae) of Palaearctics*. National Academy of Sciences of Ukraine: Kiev, Ukraine: Izdatel'stvo Naukova Dumka (in Russian), 1995.

CHAPTER 19

DISEASE MANAGEMENT OF NONLEGUMINOUS SEASONAL FORAGES

BIRESWAR SINHA*, H. CHANDRAJINI DEVI, and W. TAMPAKLEIMA CHANU

Department of Plant Pathology, College of Agriculture, CAU, Imphal, India

Corresponding author. E-mail: bireswarsinha@gmail.com

ABSTRACT

The common nonleguminous seasonal fodder crops which are widely grown in India are oat, maize, bajra, and sorghum. Like other staple food crops such as rice and wheat the production of fodder crop also suffered from various biotic and abiotic factors. Among the biotic factors, diseases caused maximum yield loss. The important fungal diseases includes covered smut, lose smut, stem rust of oat, green year, rust, ergot and smut of bajra, charcoal rot, downy mildew and leaf blight of maize, and in sorghum important diseases are downy mildew, red rot, and leaf blight.

19.1 INTRODUCTION

Fodder crops are the important sources of nutrient for the livestock animals. The common nonleguminous fodder crops include oat, maize, sorghum, and bajra. The production of the fodder crops also suffered from various constraints among which diseases and pest are one of the important factors. The present chapter will give an idea on the important diseases of some nonleguminous fodder crops their symptoms, pathogen involved, and management practices of them.

19.2 DISEASES OF OATS

Oat (*Avena sativa* L.) is of minor importance in India; being grown only in some parts of Uttar Pradesh and Punjab and some limited areas in the northern states of India. The important diseases that affect the crop are as follows.

19.2.1 COVERED SMUT

19.2.1.1 CAUSAL ORGANISM

Ustilago kolleri Wille. (Syn: *U hordei* pers.) Langerh. *U. kolleri* infects *Avena* species. Species of *Bromus*, *Hordeum*, and *Agropyron* may also be infected. In favorable conditions the covered smut may infect 10% or more plants. In this case, the damage consisting of both direct losses and losses related to depression of outwardly healthy plants may be appreciable. However, it is not now possible to consider it as an essentially dangerous disease because of the small geographic range of the covered smut distribution.

19.2.1.2 SYMPTOMS OF DAMAGE

Disease is apparent as inflorescence emerges. The plant may be slightly stunted having hard, compact, upright, and smutted panicles. The smut balls are covered with a membrane that remains intact. Sometimes narrow strips of sporiferous fungus are observed on upper leaves.

19.2.1.3 DISEASE CYCLE

The life cycle and the symptoms caused by *U. kolleri* are similar to those of *U. hordei* (pathogen of covered smut of barley). Teliospores from smutted panicles may be released by harvesting operations and mixed with oat seeds in a harvester. The infection is kept under seed coat as spores, gemmae, or dormant mycelium. Germination of teliospores results in the infection of the inner seed coat. Infection of the seedling by the dormant fungus structures occurs by penetration into the seedling axis. Then the pathogen develops in the meristem of the oat host, ultimately invading and replacing the tissues of all flowers of infected panicles. Teliospores are spherical, deep-brown, 4.6–8.1 μm in diameter, and always smooth, in contrast to *U. avenae* spores. Teliospore germination and seed infection starts at temperatures of 6–10°C. *U. kolleri* is less heat tolerant than *U. avenae*. Temperatures 25°C and humidity 100% are optimal for infection structured germination

19.2.1.4 MANAGEMENT

1. Seed treatment carboxin (vitavax @ 2.5 g/kg of seed) before sowing.
2. The smutted ears are rogue out and destroyed them.
3. Grow the resistant variety such as Brunker.

19.2.2 LOOSE SMUT OF OATS

19.2.2.1 CAUSAL ORGANISM

The causal organism of this disease is *Ustilago avenae* and *U. kolleri* are pathogens of oats causing loose smut and covered smut, respectively. Both fungi have similar life cycles, but symptoms differ. Resistance of a variety may be of short duration because of the development of new virulent races.

19.2.2.2 DISEASE CYCLE

Spores released from infected heads are blown to flowering neighbors where they lodge and remain dormant until planting or germinate and grow into the seed coat then remain dormant until planting. After planting the fungus grows within the plant and proceeds to the florets, destroying developing flowers and replacing grain with the spore mass.

19.2.2.3 SYMPTOMS OF DAMAGE

Black, powdery masses of teliospores that replace the flowers in the spring are the first signs of disease. Leaves adjacent to these flowers become black in appearance as teliospores are deposited. The bare central axis (rachis) of the panicle is all that remains, once the teliospores have been released. The outer glumes of the florets of plants with covered smut remain intact and persist as a white–grey, parchment-like covering over the jet-black teliospores. Symptoms are generally first noticed as depleted grain sites at the grain ripening stages due to their replacement by the smut spores. Smutted oat panicles sometimes fail to emerge properly from the sheath. The leaf blade may also show smut pustules. Smutted plants are shorter than the healthy ones. It is seed-borne disease.

19.2.2.4 MANAGEMENT

Seed treatments with systemic fungicides mixtures such as fuberidazole + triadimenol give effective control. Crop inspection and certification of seed crops reduces disease incidence. The bacterium *Pseudomonas chlororaphis* is reported as offering effective control with no toxicological effects. Since certified seed is grown in inspected fields and usually is treated with a fungicide, the use of certified seed is recommended. Seed treatment with fungicides for smut control is also recommended.

19.2.3 STEM RUST

Causal organism is *Puccinia graminis* f. sp. *avenae* Eriksson and E. Henning.

19.2.3.1 SYMPTOMS OF DAMAGE

The disease appears as elongated reddish–brown pustules mainly on stems but also on leaves and heads. The powdery spore masses in the pustules can dislodge readily. Stem rust causes yield losses through absorbing of nutrients that would otherwise be used for grain development, and interferes with plant vascular tissue which can lead to shriveled grain, and it can weaken the stem causing lodging.

The rust fungus causing stem rust is called *Puccinia graminis* f. sp. *avenae* Eriksson and E. Henning. It produces two types of spores on the oat plant: the reddish–brown urediniospores and the black teliospores. The pustule containing the spores is called a sorus. Urediospores develop in a uredium while teliospores develop in a telium. Uredinia develop on leaves, sheaths, and panicles of oats. They are elongated, often surrounded by the ruptured epidermis, and contain a mass of reddish–brown powdery urediniospores. Microscopically, the urediniospores are dehiscent, thick-walled, and covered with spines. They are elliptical and about 20 μm × 30 μm (15–24 μm × 21–40 μm). The teliospores develop late in the season in the pustules, which turn black. Microscopically, they are two-celled, thick-walled (with up to five-wall layers), and are thickened at the apical end, 15–20 μm × 40–60 μm and lack the "crown" seen on leaf rust teliospores (Rangaswami and Mahadevan, 2010).

19.2.3.2 MANAGEMENT

Spray zineb or maneb at the rate of 2 g/L of water in case of heavy infestation. Eradicate common barberry, as the sexual stage occurs on this alternate host, giving rise to new races.

19.2.4 CROWN RUST

The rust fungus causing leaf rust is called *Puccinia coronate* var *avenae* Schumach. It produces two types of spores on the oat plant: the yellow–orange urediniospores and the black teliospores. The pustule containing the spores is called a sorus. Urediospores develop in a uredinium while teliospores develop in a telium. Uredinia develop on leaves, sheaths, and panicles of oats. These pustules are oblong and yellow–orange in color. Microscopically, the urediniospores are spherical to ovate, orange–yellow, markedly echinulate with four germ pores located irregularly. Telia develop late in the season as leaves age as black pustules that are covered by the epidermis, except when they form in uredinia. They contain the diagnostic teliospores, which are black in mass. Microscopically, the two-celled brown teliospores are constricted slightly at the septum; the apex is thickened and has several blunt processes that form a crown-like apex.

19.2.4.1 SYMPTOMS OF DAMAGE

The characteristic symptom is the development of round to oblong, orange to yellow pustules, primarily on leaves but also on stems and heads. The powdery spore masses in the pustules are readily dislodged. The pustule areas turn black with age.

Losses result from damage to leaves (particularly the flag leaf), which leads to reduced photosynthesis and transport of carbohydrates to the developing grain. This causes shriveled grain and reduced grain quality. There is not much to be done at this stage of the growing season if rust is found. However, in future growing season's control options would include planting resistant varieties, seeding early if possible, and application of fungicides (Agrios, 1978).

19.2.4.2 MANAGEMENT

Oat crown rust races have been shifting or changing rapidly and resistance to the prevalent races has been difficult to maintain. In 1996, only Paul and Milton were designated as resistant to the prevalent races of oat crown rust. Check current variety circulars for descriptions of variety tolerance to crown rust. If possible, plant early.

Control wild oats; they serve as a susceptible host in early spring for spores from buckthorn. Eradicate wild buckthorn within a mile of oat fields; the fungus completes the sexual stage on this plant and new races can develop from the sexual process. Mancozeb fungicides (various trade names) and Tilt foliar fungicides are registered for oat and could be controlled the disease.

19.2.5 LEAF BLOTCH

19.2.5.1 CAUSAL ORGANISM

This disease is also known as Helminthosporium (red–brown) leaf spot is caused by the fungus *Drechslera avenae*, asexual stage of *Pyrenophora avenae*. It is a common disease of oats and may occasionally attack species of *Hordeum*, *Koeleria*, and *Arrhenatherium*. Sexual stage *P. avenae* form in the spring on crop debris. Disease attacks leaves and kernels of oats. Oblong to elongate, reddish–brown spots may appear on seedling leaves soon after emergence. Spots on stems may be long and narrow or broad and irregular in shape. At start, spots are little white in the center, bordered by red brown zone, 1–3 mm × 1–2 mm. Later spots are elliptic, gray or brown in the center with red–brown border and brown zone, 20–40 mm × 5–10 mm. The outer borders of the spots are poorly defined, brown color merging into yellow or reddish shades. Spots frequently spread over most of the infected leaf.

Oat heads may also become infected. The fungus enters the hulls surrounding the kernels and may even penetrate the kernels slightly. Diseased kernels turn brown at the basal stem. The fungus survives as spores or mycelium on plant debris and as spores on the seed, or as mycelium beneath the seed coat. The fungus spreads as spores from plant debris. Infection on plant parts is commonly associated with splashes of rain drops. Spores are often lodged in the hulls of the grain, providing a source for infection. Spores produced on the spot are dispersed by various means to other leaves of the same plant or to other plants. Spores (conidia) normally are single, straight, cylindrical, rarely inversely clavate, and pale-brown to olive-brown, with 1–9 partitions 50–130 × 15–20.

19.2.5.2 MANAGEMENT

1. Treat the seed with captan or Thiram @ 3 g/kg seed before sowing.
2. Rotate oat with other crop.

19.2.6 SEEDLING BLIGHTS

19.2.6.1 CAUSAL ORGANISM

Drechslera avenae (*Helminthosporium avenae*)

19.2.6.2 SYMPTOMS

Small, brown to black, elliptical spots develop on leaves, leaf sheaths, the lower portions of the culms, crowns, and roots of infected seedlings. These lesions are inconspicuous during the initial stages of development and generally are not noticed. The continued enlargement of these spots and the increase in numbers of spots result in a yellow discoloration of diseased plants. Many of these leaves will die, imparting an aspect of irregular areas of yellow plants. The continued growth of the pathogen in the roots or the lower stem results in root or culm rot, and finally, in the death of the seedling. Seedlings may recover if infections are limited to the leaves and leaf sheaths. These plants will be stunted and the leaf infections may persist throughout the winter. Conidiospores produced on these lesions will infect new leaves, thus perpetuating the disease into the spring.

19.2.6.3 MANAGEMENT

Treat seed with captan or Thiram @ 3 g/kg of seed before sowing. Plowing of plant debris hastens decomposition and consequently, will contribute to the elimination of the pathogen in the soil. Rotation of oats to barley, rye, or wheat, and vice versa, for 2 years is another control measure, since *Helminthosporium avenae* will attack oats but not barley, wheat, or rye.

19.2.7 RED LEAF OF OATS

It is a virus disease caused by *cereal yellow dwarf virus*. This virus is transmitted by *Rhopalosiphum* spp. Pale blotches develop on the virus infected leaf tips. Later on, due to fusion of the blotches, the leaf becomes reddish-brown. In advanced stages, the entire lamina becomes discolored, plant stunted and bushy in appearance.

19.3 DISEASES OF BAJRA OR PEARL MILLET *(Pennisetum typhoides)*

19.3.1 DOWNY MILDEW OR GREEN EAR

Downy mildew is the most common disease in all areas where bajra is grown. It occurs in many parts of Africa, as well as in India, where it was first reported by Butler in 1907. Disease is severe in ill drained and low lying areas. Losses due to the disease may be as high as 30–45% in the high yielding varieties. The disease occurred in epidemic form in 1970 and 1983 devastating the popular hybrids, namely, HB 3, and BJ 104.

19.3.1.1 SYMPTOMS OF DAMAGE

Infection is mainly systemic and symptoms appear on the leaves and the earhead. The first symptoms can appear in seedlings at three to four leaf stages. The affected leaves show patches of light green to light yellow color on the upper surface of leaves and the corresponding lower surface bears white downy growth of the fungus. The downy growth seen on infected leaves consists of sporangiophores and sporangia. The yellow discoloration often turns to streaks along veins. The infected plants tiller excessively and are dwarfed. As the disease advances, the streaks turn brown and the leaves shred at the tips only. In affected plants, ears fail to form or if formed, they are completely or partially malformed into twisted green leafy structures; hence the name green ear disease. The infection converts the various floral parts, including glumes, palea, stamens, and pistil into green, linear leafy structures of variable length. As the disease advances, the green leafy structures become brown and dry bearing masses of oospores (Thakur and King, 1988).

19.3.1.2 CAUSAL ORGANISM

The causal organism of this disease is *Sclerospora graminicola*. The downy mildew fungus develops systematically in the host plant. The mycelium is systemic, nonseptate and intercellular in the parenchymatous tissues sending branched haustoria into the host cells. Short, stout, hyaline sporangiophores arise through stomata and branch irregularly to produce sterigmata bearing the sporangia. Sporangia are hyaline, thin walled and elliptical, and bear prominent papilla. Oospores are round in shape, surrounded by a smooth, thick, and yellowish-brown wall. They germinate in water, releasing 3–12 zoospores.

19.3.1.3 DISEASE CYCLE

The oospores remain viable in soil for 5 years or longer giving rise to the primary infection on the host seedling, which takes place by direct penetration of root hairs and the coleoptiles by the germ tubes. Oospores attached to the seed also cause primary and systemic infection of seedlings. Secondary spread is through sporangia, which are active during rainy season, disseminated by air and water. Secondary infection may not develop into systemic infection but leads to local infection. The pathogen readily infects teosinte (*Euchlaena mexicana*) and *Setaria italica*. Formation of sporangiophores and sporangia is favored by very high humidity (90%), presence of water on the leaves, and low temperature of 15–25°C. Light drizzling accompanied by cool weather is highly favorable.

19.3.1.4 MANAGEMENT

Selection of seed from healthy crop. Collect diseased plants, especially before oospores are formed, and burn them; deep plow in summer; and rogue out infected plants. Prolonged crop rotation would be beneficial. Grow resistant verities such as WCC 75, PHB 10, ICMH 451, ICTP 8203, Mallikarjuna, HB-1, HB 5, and PHB 14. Grow tolerant varieties such as MBH 118, CM 46, Balaji composite, Nagarjuna composite, Visakha composite, New Vijaya composite, RBS 2, etc. Treat the seeds with metalaxyl (Apron 35 SD) @ 6 g/kg or Thiram or captan @ 4 g/kg. Spray mancozeb @ 0.25% or metalaxyl (ridomil MZ) @ 0.2% starting from 30 days after sowing in the field.

19.3.2 RUST OF BAJRA

Rust is a serious disease of bajra, occurring wherever the crop is grown.

19.3.2.1 SYMPTOMS OF DAMAGE

Symptoms first appear mostly on lower leaves as minute, round raised reddish-brown pustules. Uredosori occur in groups on both surfaces of leaf and leaf sheath. The pustules may also be formed on stem and peduncles. Dark brown to black teliospores are produced late in the season in the uredosori or teleutosori. In case of severe infections, whole leaf may wither completely presenting a scorched appearance to the field.

19.3.2.2 CAUSAL ORGANISM

It is caused by *Puccinia penniseti.* The rust is heteroecious. The fungus has a long life cycle producing uredial and telial stages on bajra and aecial and pycnial stages on several species of *Solanum*, including brinjal (*Solanum melongena*). Uredospores are oval, elliptic or pyriform with four germ pores, sparsely echinulated and pedicellate. Teliospores are dark brown in color, two- celled, cylindrical to club shaped, apex flattered, broad at top, and tapering toward base. The teliospores germinate to produce the characteristics three-septate, four-celled promycelium, and four basidiospores, one from each cell. The basidiospores are capable of infecting at least five species of *Solanum*, including *S. melongena*, *S. torvum*, *S. pubescens*, *S. xanthocarpum*, and *S. panduriforme*. Closer spacing, presence of abundant brinjal plants and other species of *Solanum*. The uredial stage also occurs on the species of *Pennisetum*, including *P. leonis*, *P. purpureum*, *P. orientale*, *P. spicatum* and *P. polystachyon.*

19.3.2.3 DISEASE CYCLE

This rust repeats the uredial cycle on bajra, but the teliospore stage must pass over to the alternate host to complete the life cycle. Primary infection is from the alternate host, brinjal, in nature. Secondary spread is through wind-borne uredospores. The uredial stages also occur on several species of *Pennisetum*. Low temperature of 10–12°C favors teliospore germination. A spell of rainy weather favors the onset of the disease.

19.3.2.4 MANAGEMENT

Removal and destruction of alternate hosts. Spray thrice at 15 days interval with wettable sulfur @ 0.3% starting from 21 days after planting. Grow resistant varieties such as RT 814-3, PT 826/4, PT 829/5, etc.

19.3.3 ERGOT OR SUGARY DISEASE

During 1967–1978, the disease broke out in epiphytotic proportions on newly introduced hybrid bajra varieties. On HB-1 and HB-2 hybrids, the disease occurred in epidemic form and caused 25% losses in grain yield in Bagalkot, Belgaum, and Bijapur areas of Karnataka. In severe infections, 41–70% yield losses are also reported (Saha and Dhaliwal, 2006).

19.3.3.1 SYMPTOMS OF DAMAGE

The disease occurs only at the time of flowering. Small droplets of a light, honey-colored dew-like substance exude from infected spikelets. Under severe infection many such spikelets exude plenty of honeydew which trickles along the earhead onto the upper leaves making them sticky. This attracts several insects. In the later stages, the infected ovary turns into small dark brown sclerotial bodies larger than the seed and with a pointed apex which protrude from the florets in place of grain. The fungus attacks the ovary and grows profusely producing masses of hyphae which form sclerotial bodies.

19.3.3.2 CAUSAL ORGANISM

The causal organism is *Claviceps fusiformis* or *C. microcephala.* The fungus attacks the ovary and grows profusely producing masses of hyphae which form the sclerotium. The pathogen produces septate mycelium which produces conidiophores which are closely arranged. Conidia are hyaline and single-celled. The sclerotia are small and dark grey but white inside. Sclerotia are 3–8-mm long and 0.3–15-mm broad.

19.3.3.3 DISEASE CYCLE

The fungus spreads from plant to plant in a conidial stage. The primary infection takes place by germinating sclerotia present in the soil. Secondary spread is by insects or air-borne conidia and ascospores. The role of collateral hosts such as *Cenchrus ciliaris* and *C. setigerus* in perpetuation of fungus is significant. The fungus also infects other species of *Pennisetum.* Flowers are susceptible to the infection only after stigma emergence and before pollination and fertilization. High relative humidity (80%) with a temperature of 20–30°C during flowering period, favor the disease development.

19.3.4 SMUT OF BAJRA

Smut is one of the major panicle diseases of pearl millet, caused by *Tolyposporium penicillariae* Bref. The disease has occurred in almost all the areas. The major outbreak of this disease in recent years has proved its economic importance as a serious threat to pearl millet production in northern India, particularly in Rajasthan, Haryana, Punjab, and Gujarat due to commercial cultivation of F1 hybrids (Thakur and King, 1988) where pearl millet is cultivated.

19.3.4.1 SYMPTOMS OF DAMAGE

Symptoms of the disease become apparent at the time of grain setting. The pathogen infects few florets and transforms them into large oval-shaped sacs (sori) containing black powder (smut spores). Initially the sori are larger and greener than normal healthy grains and when the sori mature they become dark brown and are easily broken and release millions of black smut spore balls.

19.3.4.2 CAUSAL ORGANISM

The Causal organism is *Tolyposporium penicillariae* Bref. The fungus infects developing flowers and the mycelium aggravates in the ovary and rounds off into chlamydospores. Meanwhile, a wall partly of host and partly of fungus tissues forms into a sorus. The fungus is mostly confined to the sorus. The sori contain spores which are usually in balls and are not easy to separate. No columella is present. Each spore is angular to round and light brown colored with a rough wall. The spores germinate to produce four-celled promycelium on which the sporidia are formed.

19.3.4.3 DISEASE CYCLE

The primary source of inoculum is spore balls in the soil from the previously infected crop along with externally contaminated seed used for sowing. The spore balls germinate to produce dense mycelial network at favorable temperature and relative humidity. The germinating teliospores form promycelia and sporidia, which infect the host at flowering. After the formation of dikaryotic infection, hypha infection takes place through young emerging stigmas. Spore balls are released from mature sori, which under favorable weather conditions germinate to produce another crop of sporidia. Optimum environmental conditions for maximum infection include: temperatures between 25°C and 35°C and slightly acidic soils favor the disease development

19.3.4.4 MANAGEMENT

Removal and destruction of affected earheads. Seed treatment with Thiram or captan @ 3 g/kg seed is required. Spray with captatol, zineb, and heptene were also reported to be effective in controlling smut in pearl millet. Grow resistant varieties such as, ICMS 8283, ICMV 1, ICMV 82132, DC 7, MPP 7131, and MPP 7108. Spray carboxin or zineb @ 0.2%.

19.4 DISEASES OF MAIZE

19.4.1 CHARCOAL ROT

19.4.1.1 SYMPTOMS

The affected plants exhibit wilting symptoms. The stalk of the infected plants can be recognized by grayish streak. The pith becomes shredded and grayish black minute sclerotia develop on the vascular bundles. Shredding of the interior of the stalk often causes stalks to break in the region of the crown. The crown region of the infected plant becomes dark in color. Shredding of root bark and disintegration of root system are the common features. Split open stalks have numerous black sclerotia on vascular strands, giving the interior of the stalks a charred appearance which is a characteristic symptom of the disease. Sclerotia may also be found on the roots.

19.4.1.2 PATHOGEN

The disease is caused by *Macrophomina phaseolina* (*Rhizoctonia bataticola*). The fungus produces large number of sclerotia which are round and black in color. Sometimes, it produces pycnidia on the stems or stalks.

19.4.1.3 FAVORABLE CONDITIONS

Favorable conditions are high temperature (37°C) at the time of silking and low soil moisture (drought). Imbalanced fertilizer application and high plant density influence disease prevalence and severity.

19.4.1.4 DISEASE CYCLE

The fungus has a wide host range, attacking sorghum, pearl millet, finger millet, and pulses. It survives for more than 16 years in the infected plant debris. The primary source of infection is through soil-borne sclerotia. The pathogen also attacks many other hosts, which helps in its perpetuation. Since the fungus is a facultative parasite it is capable of living saprophytically on dead organic tissues, particularly many of its natural hosts producing sclerotial bodies. The fungus over winters as a sclerotia in the soil and infects the host at susceptible crop stage through roots and proceeds toward stem.

19.4.1.5 MANAGEMENT

- Long crop rotation with crops that are not natural host of the fungus.
- Field sanitation
- Irrigate the crops at the time of earhead emergence to maturity.
- Treat the seeds with carbendazim or captan at 2 g/kg.
- Grow disease tolerant varieties, namely, SN-65, SWS-8029, Diva, and Zenit.

19.4.2 DOWNEY MILDEW

19.4.2.1 SYMPTOMS

The most characteristic symptom is the development of chlorotic streaks on the leaves. Plants exhibit a stunted and bushy appearance due to shortening of the internodes. White downy growth is seen on the lower surface of leaf. Downy growth also occurs on bracts of green unopened male flowers in the tassel. Small to large leaves are noticed in the tassel. Proliferation of axillary buds on the stalk of tassel and the cobs is common (Crazy top).

19.4.2.2 PATHOGEN

The pathogens associated with this disease are: *Peronosclerospora sorghi*, *Peronosclerospora philippinensis*, *P. sorghi*, *Peronosclerospora sacchari*, *Peronosclerospora maydis*, and *Sclerophthora graminicola*, *S. rayssiaie*. The fungus grows as white downy growth on both surface of the leaves, consisting of sporangiophores and sporangia. Sporangiophores are quite short and stout, branch profusely into series of pointed sterigmata which bear hyaline, oblong or ovoid sporangia (conidia). Sporangia germinate directly and infect the plants. In advanced stages, oospores are formed which are spherical, thick walled, and deep brown.

19.4.2.3 FAVORABLE CONDITIONS

- Low temperature (21–33°C)
- High relative humidity (90%) and drizzling
- Water logging condition
- Young plants are highly susceptible.

19.4.2.4 DISEASE CYCLE

The primary source of infection is through oospores in soil and also dormant mycelium present in the infected maize seeds. Secondary spread is through air-borne conidia. Crazy top is caused by a soil-borne fungus, which overwinters as oospores either within infected tissue or in the soil. During periods of flooding, the oospores germinate and swimming zoospores are produced. The zoospores infect the growing point of the young corn plants.

19.4.2.5 MANAGEMENT

- Deep summer plowing
- Destruction of plant debris
- Removal and destruction of collateral hosts
- Crop rotation with pulses
- Rogue out infected plants

- Treat the seeds with Metalaxyl at 6 g/kg
- Spray the crop 3–4 times with Metalaxyl + Mancozeb @ 0.2% starting from 20th day after sowing.
- Grow resistant varieties and hybrids, namely, DHM-1, DHM-103, DMR-5, and Ganaga II.

19.4.3 LEAF BLIGHT

19.4.3.1 SYMPTOMS

The fungus affects the crop at young stage. Small, yellowish, and round to oval spots are seen on the leaves. The spots gradually increase in area into bigger elliptical spots and are straw to grayish-brown in the centre with dark brown margins; the spots coalesce giving blighted appearance. The surface is covered with olive green velvety masses of conidia and conidiophores.

19.4.3.2 PATHOGEN

The disease is caused by *Helminthosporium maydis* (Syn: *H. turcicum*). Conidiophores are in group, geniculate, mid dark brown, pale near the apex and smooth. Conidia are distinctly curved, fusiform, and pale to mid dark golden brown with 5–11 psuedosepta.

19.4.3.3 FAVORABLE CONDITIONS

Optimum temperature for the germination of conidia is 8–27°C provided with free water on the leaf. Infection takes place early in the wet season.

19.4.3.4 DISEASE CYCLE

It is a seed-borne fungus. It also infects sorghum, wheat, barley, oats, sugarcane, and spores of the fungus are also found to associate with seeds of green gram, black gram, cowpea, varagu, Sudan grass, Johnson grass, and teosinte.

19.4.3.5 MANAGEMENT

- Crop rotation
- Grow resistant hybrids like DHM-1
- Treat the seeds with captan or Thiram at 4 g/kg
- Spray mancozeb 2 kg or captan 1 kg/ha.

19.4.4 RUST

19.4.4.1 SYMPTOM

Circular to oval, elongated cinnamon–brown powdery pustules are scattered over both surface of the leaves. As the plant matures, the pustules become brown to black owing to the replacement of red uredospores by black teliospores.

19.4.4.2 PATHOGEN

The disease is caused by *Puccinia sorghi*. Uredospores are globose or elliptical finely echinulate, yellowish-brown with four germ pores. Teliospores are brownish-black, or dark brown, oblong to ellipsoidal, and rounded to flattened at the apex. They are two-celled and slightly constricted at the septum and the spore wall is thickened at the apex.

19.4.4.3 FAVORABLE CONDITIONS

Favorable conditions are cool temperature and high relative humidity.

19.4.4.4 DISEASE CYCLE

Primary source of inoculums is uredospores surviving on alternate hosts, namely, *Oxalis corniculata* and *Euchlaena mexicana*

19.4.4.5 MANAGEMENT

- Remove the alternate hosts
- Spray mancozeb at 2 kg/ha

19.4.5 HEAD SMUT

19.4.5.1 SYMPTOMS

Symptoms are usually noticed on the cob and tassel. Large smut sori replace the tassel and the ear. Sometimes the tassel is partially or wholly converted into smut sorus. The smutted plants are stunted produce little yield and remain greener than that of the rest of the plants.

19.4.5.2 PATHOGEN

It is caused by *Sphacelotheca reiliana*. Smut spores are produced in large numbers which are reddish brown to black, thick-walled, finely spined, and spherical.

19.4.5.3 FAVORABLE CONDITIONS

Low temperature favors more infection and this fungus also infects the sorghum.

19.4.5.4 DISEASE CYCLE

The smut spores retain its viability for 2 years. The fungus is externally seed-borne and soil-borne. The major source of infection is through soil-borne chlamydospores.

19.4.5.5 MANAGEMENT

- Field sanitation
- Crop rotation with pulses

- Seed treatment: Systemic fungicides such as carboxin, vitavax, and benlate @ 2.0 g/kg seed and Tilt (propioconazole) 25 EC @ 0.1%
- Use resistant cultivars or at least avoid very susceptible cultivars

19.4.6 BANDED LEAF AND SHEATH BLIGHT

19.4.6.1 SYMPTOMS

The disease is caused by *Rhizoctonia solani* f. sp. *sasakii.* Characteristic symptoms include concentric bands and rings on infected leaves and sheaths that are discolored, brown, tan, or grey in color. The disease develops on leaves and sheaths and can spread to the ears causing ear rot.

19.4.6.2 FAVORABLE CONDITIONS

Favorable conditions are warm humid conditions, water logged conditions, high crop densities impact disease severity, and hybrids are much prone to the disease.

19.4.6.3 DISEASE CYCLE

Rhizoctonia solani survives in the soil and on infected crop debris as sclerotia or mycelium. Sclerotia are known to survive for several years in the soil. The fungi spread by water (flooding), irrigation, movement of contaminated soil, and plant debris. At the onset of the growing season, in response to favorable humidity and temperatures (15–35°C), fungal growth is attracted to freshly planted host crops by chemical stimulants released by growing plant cells. The fungi infect plants, leading to characteristic symptoms on the stem, sheaths, leaves, and ears. The fungi overwinter as sclerotia or in infected crop debris.

19.4.6.4 MANAGEMENT

- Lack of resistant commercial varieties
- Available tolerant germplasm should be cultivated

- *Trichoderma* bioformulations—through seed dressing and soil application
- Stripping of 2–3 lower leaves
- Composting of hardwood
- Fields should be well drained prior to planting
- Seeds should be planted on raised beds

19.4.7 BACTERIAL STALK ROT

19.4.7.1 SYMPTOMS

The basal internodes develop soft rot and give a water soaked appearance. A mild sweet fermenting odor accompanies such rotting. Leaves some time show signs of wilting and affected plants topple down in few days. Ears and shank may also show rot. They fail to develop further and the ears hang down simply from the plant.

19.4.7.2 PATHOGEN

Erwinia dissolvens is a motile, Gram-negative, rod-shaped bacterium.

19.4.7.3 FAVORABLE CONDITIONS

Favorable conditions are high temperature (88–95°F), high amounts of rainfall and flooding, overhead irrigation, and insect injury.

19.4.7.4 DISEASE CYCLE

Borer insects play a significant role in initiation of the disease. The organism is soil-borne and makes its entry through wounds and injuries on the host surface. The organism survives saprophytically on debris of infected materials and serves primary inoculum in the next season.

19.4.7.5 MANAGEMENT

- Use of disease resistance varieties, that is, Hybrids Ganga Safed-2, DHM 103
- Sanitation—removal of infected crop residues
- Avoid water logging and poor drainage
- Avoid excessive irrigation
- Avoid injury to plant parts
- Three application of bleaching powder (10%) @ 16.5 kg/ha at the time of sowing, earthling up and tasseling stage.

19.4.8 BROWN SPOT

Brown spot of maize was first reported in India in 1910. Subsequently, it has been reported from other countries of the world, such as, China, Japan and, the United States. It is a sporadic disease and minor disease, but may become severe, when environmental conditions are favorable for disease occurrence.

19.4.8.1 SYMPTOMS

The disease attacks the leaf blades, leaf sheaths and culms of young plants. Water-soaked lesions, which are oval, later turn into light green and finally brown.

19.4.8.2 PATHOGEN

Physoderma maydis is an obligate parasite. The older lesions contain a large number of thick-walled resting spores or resting sporangia. They are smooth, brown and flattened on one side, where a cap-like lid is present. The resting sporangia survive in the diseased plant debris or in the soil and carry over the disease from one season to the next season.

19.4.8.3 MODE OF SURVIVAL, SPREAD, AND EPIDEMIOLOGY

The resting sporangia can survive in the crop remains and debris for up to 4 years in the field. Primary infection occurs through the soil-borne resting sporangia and secondary dissemination is by spore dust carried by wind. The disease is common in low lying and ill drained fields. High temperatures of 28–29°C and abundant moisture during the early phase of crop growth are favorable for the occurrence of the disease.

19.4.8.4 MANAGEMENT

- Field sanitation
- Crop rotation reduce the disease intensity
- No chemical control measures have been suggested to control the disease
- Use of tolerant varieties.

19.4.9 MOSAIC: MAIZE MOSAIC POTYVIRUS

19.4.9.1 SYMPTOMS

Symptoms often begin as chlorotic spots and streaks on green, young leaves, which later develop into a mottle or a mosaic pattern. Viral strain, corn genotype, and stage of corn development at the time of infection will affect the type of symptoms. Upper internodes of corn may be shortened, and excessive tillering may occur. Ear formation and development may slow, which may cause grain yield loss. Hybrids infected early in their growth stage may be stunted.

19.4.9.2 PATHOGEN

It is caused by *Maize mosaic potyvirus* (MSV). Virions are flexuous, 750–900-nm long, ssRNA genome. This pathogen consists of four strains. Aphids = (15 species) can transmit MDMV nonpersistently. It can be seed-transmitted at a low frequency or mechanically transmitted by leaf rubbing, etc. It infects Johnson grass and sorghum.

19.4.9.3 TIME OF OCCURRENCE

Symptoms appear 6 weeks after aphids feed and transmit this virus.

19.4.9.4 TRANSMISSION

It is transmitted in nature by leafhopper vector, *Peregrinus maidis*

19.4.9.5 FAVORABLE CONDITION

Average to warm temperatures. Nearby Johnson grass infected with MDMV may increase disease.

19.4.9.6 MANAGEMENT

- Use tolerant commercial corn hybrids
- Avoidance of aphid vectors
- Control of Johnson grass with herbicides may be beneficial.

19.4.10 MAIZE STREAK VIRUS: GENUS *Mastrevirus*, FAMILY GEMINIVIRIDAE

19.4.10.1 SYMPTOMS

The virus causes a white to yellowish streaking on the leaves. The streaks are very narrow, more or less broken and run parallel along the leaves. Eventually, the leaves turn yellow with long lines of green patches. Plants infected at early stage usually do not produce any cobs.

19.4.10.2 PATHOGEN

MSV and all related grass master viruses have single-component, circular, single-stranded DNA genomes of approximately 2700 bases, encapsidated in 22 nm × 38 nm geminate particles. Particles are generally stable in buffers of pH 4–8.

19.4.10.3 TRANSMISSION

MSV is mainly transmitted by *Cicadulina mbila* , but other leafhopper species, such as *C. storeyi*, *C. arachidis*, and *C. dabrowski*, are also able to transmit the virus. The virus may be acquired in less than 1 h (minimum acquisition time 15 s) and may be inoculated in 5 min; latent period in the vector is 6–12 h at 30°C, which coincides with the first appearance of the virus in the body fluids. Following acquisition of the virus by a leafhopper, it becomes viruliferous within 30 h and is able to transmit for the rest of its life.

19.4.10.4 MANAGEMENT

- Use of tolerant/resistant varieties
- Early rouging
- Eradication of grass weeds
- Control vector by spraying with Dimethoate, Malathion
- Avoid overlap of two maize crops
- Crop rotation
- Use certified maize seed.

19.5 DISEASES OF SORGHUM

19.5.1 SORGHUM DOWNY MILDEW

19.5.1.1 SYMPTOMS

Leaves have yellow to white stripes extending from the base of the leaf toward the leaf tip. Infected leaves often will be narrower than usual, erect, and shredded. A white, downy mycelium may develop on both surfaces of leaves. Plants may be stunted, chlorotic, and have an irregular seed set. Symptoms vary according to plant tolerance

19.5.1.2 PATHOGEN: *Peronosclerospora sorghi*

The causal organism is an obligate parasite. It grows systemically in the host plant. The mycelium is coenocytic, intercellular, and is present in the parenchymatous tissues of the roots and stem and the mesophyll tissues of the leaves. The hyphae produce finger-like haustoria. Sporangiospores emerge singly or in clusters through the stomata. Sporangia are produced singly at the tips of the sterigmata. They are single-celled, thin-walled, hyaline, globose, and lack apical papilla.

19.5.1.3 DISEASE CYCLE

The primary infection is by means of oospores present in the soil which germinate and initiate the systemic infection. Oospores persist in the soil for several years. Secondary spread is by air-borne sporangia. Presence of mycelium of the fungus in the seeds of systemically infected plants is also a source of infection. The disease has been known to occur through a collateral host, *Heteropogen centortus* on which the fungus perpetuates of the host. The breakdown of tissue causes shredding. The oospores either fall to the soil or are wind-blown, often within host tissue. They can remain viable in the soil for 5–10 years. Conidia are formed at night in large numbers. The optimum temperature for production is 20–23°C.

19.5.1.4 MODE OF SURVIVAL AND SPREAD

The pathogen is both soil-borne and seed-borne. Plant debris and haystacks harbor the oospores. The oospores germinate when they come into contact with the germinating seedlings and cause primary systemic infection. Secondary spread is by means of sporangia dispersed by wind, rainwater, direct contact, and by some insects mechanically.

19.5.1.5 EPIDEMIOLOGY

Optimum temperature for sporulation and sporangial germination is 21–23°C. Formation of sporangia is suppressed at temperatures below 15°C. High relative humidity and damp, saturated atmosphere with a film of water on the leaf surface, and temperature 21°C is highly favorable for severe disease occurrence.

19.5.1.6 DISEASE MANAGEMENT

- Grow moderately resistant varieties such as Co25 and Co26
- Deep plowing
- Plant in cool soils
- Avoid rotation of corn with sorghum and adjacent planting of corn and sorghum

- Destroy shattercane, because it is an alternate host for the pathogen
- Systemic fungicide seed treatments may be helpful
- Seed dressing with Metalaxyl at 2 g/kg of seeds is found to be effective in controlling seed-borne infection.

19.5.2 ANTHRACNOSE AND RED ROT

19.5.2.1 SYMPTOMS

The fungus causes both leaf spot (anthracnose) and stalk rot (redrot). The disease appears as small red-colored spots on both surfaces of the leaf. The centre of the spot is white in color encircled by red, purple or brown margin. Numerous small black dots like acervuli are seen on the white surface of the lesions. Red rot can be characterized externally by the development of circular cankers, particularly in the inflorescence. Infected stem when split open shows discoloration, which may be continuous over a large area or more generally discontinuous giving the stem a marbled appearance.

19.5.2.2 PATHOGEN

It is caused by *Colletotrichum graminicola*. The mycelium of the fungus is localized in the spot. Acervuli with setae arise through epidermis. Conidia are hyaline, single-celled, vacuolate, and falcate in shape.

19.5.2.3 FAVORABLE CONDITIONS

Favorable conditions are continuous rain, temperature of 28–30°C, and high humidity.

19.5.2.4 DISEASE CYCLE

The disease spread by means of seed-borne and air-borne conidia and also through the infected plant debris.

19.5.2.5 MANAGEMENT

- Removal and destruction of alternate hosts in and around the fields
- Field sanitation and selection of healthy seeds
- Seed dressing with captan or Thiram at 4 g/kg of seed, 24 h prior to sowing protects the crop from seed-borne infection.
- Spray the crop with Mancozeb 2 kg/ha.

19.5.3 LEAF BLIGHT

19.5.3.1 SYMPTOMS

The pathogen also causes seed rot and seedling blight of sorghum. The disease appears as small, narrow, elongated spots in the initial stage and in due course they extend along the length of the leaf. On older plants, the typical symptoms are long elliptical necrotic lesions, straw-colored in the centre with dark margins. The straw-colored centre becomes darker during sporulation. The lesions can be several centimeters long and wide. Many lesions may develop and coalesce on the leaves, destroying large areas of leaf tissue, giving the crop a burnt appearance.

19.5.3.2 PATHOGEN

The disease is caused by *Exserohilum turcicum* (Syn: *Helminthosporium turcicum*). The mycelium is localized in the infected lesion. Conidiophores emerge through stomata and are simple, olivaceous, septate, and geniculate. Conidia are long, cylindrical, slightly curved, olivaceous brown, 3–8 septate, and thick-walled. They germinate by producing a germ tube from either of the terminal cells.

19.5.3.3 FAVORABLE CONDITIONS

Moderate temperature, high atmospheric moisture, high humidity (90%), and intermittent rainfall favors the disease occurrence and spread. The pathogen attacks maize and Johnson grass.

19.5.3.4 DISEASE CYCLE

The mycelium and conidia of the pathogen can survive in the soil in the infected plant debris for a long period of time. The pathogen is found to persist in the infected plant debris. Seed-borne conidia are responsible for seedling infection. Secondary spread is through wind-borne conidia.

19.5.3.5 MANAGEMENT

- Use disease free seeds
- Field sanitation helps to reduce the inoculum potential in the soil
- Seed dressing with captan or Thiram at 4 g/kg of seed, 24 h prior to sowing protects the crop from seed-borne infection.
- Foliar spraying with Mancozeb 500 g in 250 L of water per acre control the disease effectively.

19.5.4 RUST

19.5.4.1 SYMPTOMS

The fungus affects the crop at all stages of growth. The first symptom is small flecks on the lower leaves (purple, tan or red depending upon the cultivar). Pustules (uredosori) appear on both surfaces of leaf as purplish spots which rupture to release reddish powdery masses of uredospores. Teliospores develop later sometimes in the old uredosori or in telisori, which are darker and longer than the uredosori. The pustules may also occur on the leaf sheaths and on the stalks of inflorescence.

19.5.4.2 PATHOGEN

The disease is caused by *Puccinia purpurea*. The uredospores are pedicellate, elliptical or oval, thin-walled, echinulated, and dark brown in color. The teliospores are reddish or brown in color and two-celled, rounded at the apex with one germ pore in each cell. The teliospores germinate and produce promycelium and basidiospores. Basidiospores infect *Oxalis corniculata* (alternate host) where pycnial and aecial stages arise.

19.5.4.3 SURVIVAL AND SPREAD

The uredospores survive for a short time in soil and infected debris. Presence of alternate host helps in perpetuation of the fungus.

19.5.4.4 FAVORABLE CONDITIONS

Low temperature of 10-12°C favors teliospore germination and a spell of rainy weather favors the onset of the disease.

19.5.4.5 DISEASE CYCLE

The uredospores survive for a short time in soil and infected debris. Presence of alternate host helps in perpetuation of the fungus.

19.5.4.6 MANAGEMENT

- Remove the alternate host *Oxalis comiculata*
- Spray the crop with mancozeb at 2 kg/ha
- Growing less-susceptible or resistant cultivars such as PSH 1, SPH 837, CSV 17, etc. is the only solution.

19.5.5 ERGOT OR SUGARY DISEASE

19.5.5.1 SYMPTOMS

The disease is confined to individual spikelets. The first symptom is the secretion of honeydew from infected florets. Under favorable conditions, long, straight or curved, cream to light brown, hard sclerotia develop. Often the honeydew is colonized by *Crerebella sorghivulgaris* which gives the head a blackened appearance.

19.5.5.2 PATHOGEN

The disease is caused by *Sphacelia sorghi*. The fungus produces septate mycelium. The honeydew is a concentrated suspension of conidia, which are single-celled, hyaline, elliptic or oblong.

19.5.5.3 FAVORABLE CONDITIONS

- A period of high rainfall and high humidity during flowering season
- Cool night temperature and cloudy weather aggravate the disease.

19.5.5.4 DISEASE CYCLE

The primary source of infection is through the germination of sclerotia which release ascospores that infect the ovary. The secondary spread takes place through air- and insect-borne conidia. Rain splashes also help in spreading the disease.

19.5.5.5 MANAGEMENT

- Adjust the date of sowing so that the crop does not flower during September–October when high rainfall and high humidity favor the disease.
- Spray any one of the following fungicides, namely, mancozeb 2 kg/ha (or) carbendazim at 500 g/ha at emergence of earhead (5–10% flowering stage) followed by a spray at 50% flowering and repeat the spray after a week, if necessary.
- Crops should be inspected for ergot 10–14 days after flowering during cool, wet weather, particularly around the edges.
- Estimate infection levels of about 100 heads (an average level of less than 0.3% or 1% of infected spikelet per head should be safe, depending on the intended use of the grain).
- Spray out plants if late tillers are infected.
- Increase fan speed of headers to maximize sclerote removal during harvesting.
- Harvest heavily contaminated areas of the crop separately.

- Consider mixing ergot-contaminated sorghum with clean seed to reduce ergot sclerote levels to less than 0.3% or 0.1% depending on the intended use of the grain.

19.5.6 GRAIN SMUT/KERNEL SMUT/COVERED SMUT/SHORT SMUT

19.5.6.1 SYMPTOMS

The individual grains are replaced by smut sori. The sori are oval or cyclindrical and are covered with a tough creamy skin (peridium) which often persists unbroken up to thrashing. Ratoon crops exhibit higher incidence of disease.

19.5.6.2 PATHOGEN

The disease is caused by *Sphacelotheca sorghi*. The spores are globose to sub-globose, thick-walled, olive brown in color, apparently smooth-walled. The sporadia are spindle-shaped and they multiply by repeated budding till the protoplasm is depleted.

19.5.6.3 FAVORABLE CONDITION

The disease if favored by temperature of about 25°C and medium to low soil moisture favors infection and development of the disease.

19.5.7 LOOSE SMUT/KERNEL SMUT

19.5.7.1 SYMPTOMS

The affected plants can be detected before the ears come out. They are shorter than the healthy plants with thinner stalks and marked tillering. The ears come out much earlier than the healthy. The glumes are hypertrophied and the earhead gives a loose appearance than healthy. The sorus is covered by a thin membrane which ruptures very early, exposing the spores even as the head emerges from the sheath.

19.5.7.2 PATHOGEN

The disease is caused by *Sphacelotheca cruenta*. The loose smut of sorghum is caused by the fungus *Sphacelotheca cruenta* which forms teliospores (hibernating stage of the pathogen) and sporidia in its life cycle. Spherical or ellipsoidal, tinted, smooth teliospores are 4.8 μm × 10.8 μm, rarely to 12 μm in diameter. At sprouting they give a four-celled basidium with laterally and apically growing sporidia. The sporidia are spindle-shaped or oblong, 2.8 μm × 12.7 μm in size. Mycelium is multicellular.

19.5.7.3 FAVORABLE CONDITIONS

The fungus is capable to develop at temperatures 10–32°C. The soil temperature 18–23°C and soil humidity 15–20% are optimal for mass infection of plants during seed germination.

19.5.8 LONG SMUT

19.5.8.1 SYMPTOMS

This disease is normally restricted to a relatively a small proportion of the florets which are scattered on a head. The sori are long, more or less cylindrical, elongated, slightly curved with a relatively thick creamy-brown covering membrane (peridium). The peridium splits at the apex to release black mass of spores (spore in groups of balls) among which are found several dark brown filaments, which represent the vascular bundles of the infected ovary.

19.5.8.2 PATHOGEN

The disease is caused by *Tolyposporium ehrenbergii*. The spores are united into solid, granular balls, which are black in color and intermixed with shreds of host tissue. The spore's ball contains many spores which are rather permanently united, irregular in shape, and vary in size. They are globose to elongated, dark brown in mass. The spores found in the interior of the balls are pale in color, smooth and globose to subglobose or angular in shape.

19.5.8.3 FAVORABLE CONDITION

The optimum temperature for germination of spores is 28°C. A relative humidity of 70–80% is favorable for the occurrence of the disease.

19.5.9 HEAD SMUT

19.5.9.1 SYMPTOMS

The entire head is replaced by large sori. The sorus is covered by a whitish-grey membrane of fungal tissue, which ruptures, before the head emerges from the boot leaf to expose a mass of brown smut spores. Spores are embedded in long, thin, dark-colored filaments which are the vascular bundles of the infected head.

19.5.9.2 PATHOGEN

The disease is caused by *Sphacelotheca reiliana*. The sorus is composed of loosely united smut spores and the conductive tissue of the inflorescence. The spore mass is powdery, dark brown and is quickly dispersed to expose a tangled mass of vascular strands of the host or sometimes a single central columella. The spores are reddish-brown to black in color, finely echinulated, and spherical to irregular in shape. The spores germinate to produce a four-celled promycelium and sporadia.

19.5.9.3 FAVORABLE CONDITION

Soil temperature 21–28°C and soil moisture of 15–25% is favorable for infection of seedlings.

19.5.9.4 MANAGEMENT FOR ALL SMUTS

- Affected earheads should be collected before harvest and destroyed
- Disease free seeds should be used for sowing
- Crops sown early in the season, that is, by early June usually escape disease incidence

- Solar heat treatment
- Follow crop rotation
- Seed dressing with sulphur dust or Thiram or captan at 4 g/kg of seeds, 24 h prior to sowing, eliminates the externally seed-borne spores.
- Resistant varieties for grain smut—Nandyal and Bilichigan
- Seed treatment is not done in long smut of sorghum
- Resistant variety for long smut—Irungu cholam is found to be free from infection, as the glumes cover the floral parts completely in Tamil Nadu.

19.5.10 HEAD MOULD/GRAIN MOULD/HEAD BLIGHT

More than 32 genera of fungi were found to occur on the grains of sorghum.

19.5.10.1 SYMPTOMS

The symptoms appear as mycelial growth and fructifications of the organisms on the surface of the inflorescence and grains, which may be of different colors and textures depending upon the causal organisms. Sometimes, the fungal growth may be superficial, while in the other cases the infection may go deeper into the tissues, if rains occur during the flowering and grain filling stages, severe grain molding occurs. Symptom varies depending upon the organism involved and the degree of infection.

19.5.10.2 PATHOGEN

The most frequently occurring genera are *Fusarium*, *Curvularia*, *Alternaria*, *Aspergillus*, and *Phoma*. *Fusarium semitectum* and *F. moniliforme* develop a fluffy white or pinkish coloration. *C. lunata* colors the grain black.

19.5.10.3 FAVORABLE CONDITIONS

- Wet weather following the flowering favors grain mould development
- The longer the wet period the greater the mould development
- Compact earheads are highly susceptible.

19.5.10.4 DISEASE CYCLE

The fungi mainly spread through air-borne conidia. The fungi survive as parasites as well as saprophytes in the infected plant debris.

19.5.10.5 MANAGEMENT

- Adjust the sowing time
- Grow resistant varieties such as GMRP 4, GMRP 9, GMRP 13, and tolerant varieties such as CSV 15.
- Seed disinfestations with Thiram @ 0.3% will prevent seedling infection
- Spray any one of the following fungicides in case of intermittent rainfall during earhead emergence, a week later, and during milky stage.
- Mancozeb 1 kg/ha or captan 1 kg + aureofungin-sol 100 g/ha.

19.5.11 PHANEROGAMIC PARASITE—*Striga asiatica* AND *Striga densiflora*

It is a partial root parasite and occurs mainly in the rainfed sorghum. It is a small plant with bright green leaves, grows up to a height of 15–30 cm. The plants occur in clusters of 10–20/host plant. *S. asiatica* produces red to pink flowers while *S. densiflora* produces white flowers. Each fruit contains minute seeds in abundance which survives in the soil for several years. The root exudates of sorghum stimulate the seeds of the parasite to germinate. The parasite then slowly attaches to the root of the host by haustoria and grows below the soil surface producing underground stems and roots for about 1–2 months. The parasite grows faster and appears at the base of the plant. Severe infestation causes yellowing and wilting of the host leaves. The infected plants are stunted in growth and may die prior to seed setting.

19.5.11.1 DISEASE CYCLE

The primary infection is by seeds present in the soil.

19.5.11.2 FAVORABLE CONDITIONS

Favorable conditions are soil temperature of 35°C and soil moisture of 30%.

19.5.11.3 MANAGEMENT

- Regular weeding and intercultural operation during early stages of parasite growth
- Crop rotation with cowpea, groundnut, and sunflower
- Mixing of Ethrel with soil triggers germination of *Striga* in the absence of host
- Spray fernoxone (sodium salt of 2,4-D) or agroxone (MCPA) at 450 g/500 L of water or praquat @ 1 kg/ha.
- 1% Tetrachloro dimethyl phenoxy acetate can be used for instant killing of *Striga*, if water is in scarce.

KEYWORDS

- **oats**
- **maize**
- **sorghum**
- **bajra**
- **disease**

REFERENCES

Agrios, G. N. *Plant Pathology*; Academic Press Inc.: London, UK, 1978.

Rangaswami, G.; Mahadevan, A. *Diseases of Crop Plants in India*, 4th Ed.; PHI Learning Private Limited: New Delhi, 2010.

Saha, L. R.; Dhaliwal, G. S. *Handbook of Plant Protection*, 2nd Ed.; Kalyani Publication: Ludhiana, 2006.

Thakur, R. P.; King, S. B. *Smut Disease of Pearl Millet*. Information Bulletin No. 25. International Crops Research Institute for the Semi-arid Tropics: Patancheru, India, 1988.

CHAPTER 20

AGROFORESTRY IN FODDER AND FORAGE CROPS

KANU MURMU*

Department of Agronomy, Bidhan Chandra Krishi Viswavidyalaya, Mohanpur 741252, West Bengal, India

*E-mail: kanumurmu@gmail.com

ABSTRACT

Agroforestry focuses on the wide range of trees grown on farms and other rural areas. Among these are fertilizer trees for land regeneration, soil health, and food security; fruit trees for nutrition; fodder trees for livestock; timber and energy trees for shelter and fuel wood; medicinal trees to cure diseases; and trees for minor products, namely, gums, resins or latex products. Many of these trees are multipurpose, providing a range of benefits. Most growth in meat production in the tropics over middle of the last decade was through increased animal stocking, but notable increases in output per animal for milk and eggs were achieved in some regions, especially in East and Southeast Asia. By contrast, in sub-Saharan Africa over the same period, per capita consumption remained about the same over the region as a whole, while increases in efficiency observed in poultry and dairy farming observed elsewhere were not widely achieved.

20.1 INTRODUCTION

Trees play an important role in all terrestrial ecosystems and provide a range of products and services to rural and urban people. As natural vegetation is cut for agriculture and other types of development, the benefits that trees provide are best sustained by integrating trees into agricultural system—a practice known as agroforestry. Farmers have practiced agroforestry since ancient times. Agroforestry focuses on the wide range of trees grown on farms and other rural areas. Among these are fertilizer trees for land regeneration, soil health, and food security; fruit trees for nutrition; fodder trees for livestock; timber and energy trees for shelter and fuel wood; medicinal trees to cure diseases, and trees for minor products, namely, gums, resins or latex products. Many of these trees are multipurpose, providing a range of benefits. According to the 2001 report of the Forest Survey of India, the forest cover in the country is 675,538 sq. km, constituting 20.55% of its total geographical area. Out of this, dense forest constitutes 2.68% and open forest 7.87%. The forest cover in the hilly districts is only 38.34% compared with the desired 66% area. The National Agriculture Policy (2000) emphasized the role of agroforestry for efficient nutrient cycling, nitrogen fixation, and organic matter addition and for improving drainage and underlining the need for diversification by promoting integrated and holistic development of rainfed areas on watershed basis through involvement of community to augment biomass production through agroforestry and farm forestry. The Task Force on Greening India for Livelihood Security and Sustainable Development of Planning Commission (2001) has also recommended that for sustainable agriculture, agroforestry may be introduced over an area of 14 million hectare out of 46 million hectare irrigated areas that are degrading due to soil erosion, waterlogging, and salinization. For integrated and holistic development of rainfed areas, agroforestry is to be practiced over an area of 14 million hectare out of 96 million hectare. This all will, besides ensuring ecological and economic development provides livelihood support to about 350 million people. The practice of agroforestry can help in achieving these targets. Therefore, in the quest of optimizing productivity, the multitier system came into existence. Gap of demand and supply of forest produce in India is widening and forests are unable to fulfill the demand. Agroforestry can play an important role in filling this gap and conservation of natural resources.

Comparing 1980 with the middle of the last decade, for example, the production and per capita consumption of meat, milk and eggs increased hugely in various parts of the world, most notably in East and Southeast Asia, and especially in China (FAO, 2009). Most growth in meat production in the tropics over this period was through increased animal stocking, but notable increases in output per animal for milk and eggs were achieved in some regions, especially in East and Southeast Asia. By contrast, in sub-Saharan Africa over the same period, per capita consumption remained about the same over the region as a whole, while increase in efficiency observed in poultry and dairy farming observed elsewhere were not widely achieved (FAO, 2009). In the last few years, however, increases in incomes in East Africa (World Bank, 2014) are driving greater per capita meat and dairy consumption, as there is a strong positive correlation between incomes and the consumption of animal products (Ngigi, 2005).

Agroforestry is recognized as an important component in climate-smart agriculture (defined as agriculture that brings humankind closer to safe operating spaces across spatial and temporal scales for food systems; in the context of climate change (Neufeldt et al., 2013) for both its adaptation and mitigation roles (Thorlakson and Neufeldt, 2012). One important feature is that trees and agroforestry systems provide a wide range of products and services that can substitute for each other and in the right circumstances can be produced synergistically. Similarly, livestock-keeping diversifies rural communities' production options and is often adapted to relatively marginal environments, which can promote climate-resilience (MacOpiyo et al., 2008; Thornton and Herrero, 2008).

At the same time, livestock production globally contributes about 9% of total anthropogenic carbon dioxide (CO_2), 37% of methane (CH_4) and 65% of nitrous oxide (N_2O) emissions (FAO, 2009). Along the animal food chain, major sources of greenhouse gas (GHG) emissions include: land-use change to produce animal feed (e.g., CO_2 emissions from cutting of forest for pasture and feeds crops in Latin America); other inputs into feed production (e.g., CO_2 from fossil fuels used to manufacture chemical fertilizer for application to feed crops and to transport feeds); direct animal production (e.g., CH_4 from enteric fermentation in ruminants); and manure emissions (e.g., CH_4, N_2O, and ammonia (NH_3) during storage, application, and deposition). Ruminants such as cattle and buffalo are responsible for more GHG emissions than monogastric pigs and poultry (FAO, 2009).

The origin of agroforestry practices, that is, growing trees with food crops and grasses, is believed to have been during Vedic era (Ancient period, 1000 B.C.); the agroforestry as a science is introduced only recently. The systematic research in agroforestry geared up after the establishment of the International Council for Research in Agroforestry (ICRAF) in 1977, which was renamed in 1991 as the International Centre for Research in Agroforestry. During 2001–2002, ICRAF adopted a new brand name "World Agroforestry Centre," to more fully reflect their (ICRAF's) global reach and also their more balanced research and development agenda; however their legal name "International Centre for Research in Agroforestry" will remain unchanged. In India, organized research in agroforestry was initiated in 1983 by the establishment of All India Coordinated Research Project on Agroforestry by ICAR at 20 centers and later establishment of the National Research Centre for Agroforestry at Jhansi in 1988. At present 39 centers of agroforestry are working in the country. The process of system evolution can be still observed in the natural forests through settled agriculture, animal husbandry and organized forestry with the adoption of variety of land use practices where tree is one of the components. Agroforestry systems have been the target of scientific enquiry and analysis and thus have been defined by many in different ways.

20.2 CONCEPT OF AGROFORESTRY

Agroforestry is a collective name for land-use systems involving trees combined with crops and/or animals on the same unit of land. It combines (1) production of multiple outputs with protection of resource base; (2) places emphasis on the use of multiple indigenous trees and shrubs; (3) particularly suitable for low-input conditions and fragile environments; (4) it involves the interplay of sociocultural values more than in most other land-use systems; and (5) it is structurally and functionally more complex than monoculture.

20.3 DEFINITION

Agroforestry is any sustainable land-use system that maintains or increases total yields by combining food crops (annuals) with tree crops (perennials) and/or livestock on the same unit of land, either alternately or at the same time, using management practices that suit the social and cultural characteristics of the local people and the economic and ecological conditions of the area.

Agroforestry is a collective name for a land-use system and technology; whereby woody perennials are deliberately used on the same land management unit as agricultural crops and/or animals in some form of spatial arrangement or temporal sequence. In an agroforestry system there are both ecological and economical interactions between the various components.

It needs to be clearly understood that specifying the existence of spatial–temporal arrangements among components does not help in defining agroforestry, but its value lies in classifying agroforestry examples. Multiple cropping as opposed to multiple uses is a necessary condition to agroforestry. Production diversification is not exclusive to agroforestry and does not help in defining agroforestry. The sole existence of economical interactions among the components is not a sufficient condition to define agroforestry; biological interactions must be present. Similarly, the term significant interactions among the components cannot be used objectively in defining agroforestry, and its use should be avoided. The presence of animal is not essential to agroforestry. Agroforestry implies management of at least one plant species for forage, an annual or perennial crop production. Once appropriate time limits are imposed on the system, time sequences involving at least two plant species with at least one woody perennial must be considered agroforestry. On the basis of this analysis, the final definition of agroforestry could be:

"Agroforestry is a form of multiple cropping which satisfies three basic conditions (1) there exists at least two plant species that interact biologically, (2) at least one of the plant species is a woody perennial, and (3) at least one of the plant species is managed for forage, annual or perennial crop production."

It shows that agroforestry is a new name for a set of old practices. In simple terms agroforestry is "an efficient land-use system where trees or shrubs are grown with arable crops, seeking positive interactions in enhancing productivity on the sustainable basis."

Agroforestry combines agriculture and forestry technologies to create more integrated, diverse, productive, profitable, healthy, and sustainable land-use systems. The most important agroforestry practices are windbreaks, riparian forest buffers, alley cropping, silvopasture, and forest farming. Agroforestry is a "social forestry"—its purpose is sustainable development. Practices are focused on meeting the economic, environmental, and social needs of people on their private lands. At the farm level, agroforestry is a set of practices that provide strong economic and conservation incentives for landowner adoption. Incorporated into watersheds and landscapes, agroforestry practices help to attain community/society goals for more diverse, healthy, and sustainable land-use systems.

20.4 HISTORY OF AGROFORESTRY IN INDIA

In about 700 B.C., man changed from a system of hunting and food gathering to food production. Shifting cultivation in India is prehistoric and partly a response to agroecological conditions in the region. Horticulture as coexistent with agriculture is found to have been prevalent in India from early historic period (500 B.C. to first century) when a certain amount of share in garden crops started to have been enjoyed by the king for providing irrigation. Some stray references occur in different texts of the Vedic literature. The cultivation of date-palm, banana, pomegranate, coconut, jujube, aonla, bael, lemon, and many varieties of other fruits and requirement of livestock in agriculture, mixed economy of agriculture, and cattle breeding may be traced in protohistory Chalcolithic periods of civilization. But in India, the plant husbandry (intentional sowing or planting for production of desirable plants or plant domestication) happened to start under progressively arid climatic zone from about preNeolithic (*Ficus religiosa*), palasa (*Butea monosperma*), and varana (*Crataeva roxburghii*) in Indian folk-life has been mentioned in ancient literature of Rig Veda, Atharva Veda, and other ancient scriptures. Traditional agroforestry systems manifest rural people's knowledge and methods to benefit from complimentary uses of annuals and woody perennials on the sustained basis. It also indicates that farmers have a closer association with trees than any other social group and promoters of forests.

In Central America it has been a traditional practice for a long time for farmers to plant about two dozen species on a small piece of land configuring them in different planes. In Europe, until middle ages it was the general custom to clear fell, degraded forests, burn slash, cultivate food crops for varying periods on cleared area and plant trees before or along with, or after sowing agricultural crops. This farming system was widely practiced in Finland up to the end of the last century and in a few areas in Germany as late as the 1920s. In certain far-east countries practice; people clear forest for agricultural use—they deliberately spared certain trees which by the end of the rice-growing season provided partial canopy of new foliage to prevent excessive exposure of soil to sun.

The farmers and land owners in different parts of the country integrate a variety of woody perennials in their crop and livestock production fields, depending upon the agroclimates and local needs. Most of these practices are, however very location specific and information on these are mostly anecdotal. Therefore, their benefits have remained vastly under exploited to other potential sites. It has now been well-recognized that agroforestry can address some of the major land-use problems of rainfed and irrigated farming systems in India, and that a great deal can be accomplished by improving indigenous systems. With the current interests in agroforestry worldwide, attempts are being made in India to introduce agroforestry techniques using indigenous and exotic multipurpose and nitrogen-fixing woody perennials.

20.5 TRADITIONAL AGROFORESTRY SYSTEMS IN INDIA

Agroforestry is widespread in all ecological and geographical regions of India. The systems vary enormously in their structural complexity and species diversity, their productive and protective attributes, and their socioeconomic dimensions. They range from apparently simple forms of shifting cultivation to complex home gardens: from systems involving sparse stands of trees on farm lands (e.g., *Prosopis cineraria* khejri tree in arid regions of western India) to high-density complex multistoried homesteads of humid lowlands: from systems in which trees play a predominantly 'service' role (e.g., shelter belts) to those in which they provide main salable products (e.g., intercropping with plantation crops). Most of these are anecdotal but in some, enough research efforts have been carried out in recent times.

20.5.1 SHIFTING CULTIVATION (SLASH AND BURN SYSTEM)

It refers to farming system in north-eastern high rainfall areas in which land under natural vegetation (usually forests) is cleared by slash and burn method, cropped with common arable crops for a few years and then left unattended when natural vegetation regenerates. Traditionally the fallow period is 10–20 years but in recent times it is reduced to 2–5 years in many areas. Due to the increasing trends of population pressure, the fallow period is drastically reduced and system has degenerated causing serious soil erosion, depleting soil fertility resulting to low productivity. In north-eastern India many annual and perennial crops with diverse growth habits are being grown.

At times annual crops such as potato, rice, maize, and ginger are grown in monoculture or mixed culture along with *Pinus kesiya*. Another important attribute of the system is secondary succession of vegetation during fallow period. The tree species which may be considered suitable for afforestation of fallow areas or to intercrop with arable crops must be fast growing preferably nitrogen-fixing and must efficiently recycle available nutrients within the system shortening time required to restore fertility. These may include species of *Acacia*, *Albizia*, *Alnus*, *Casuarina*, *Erythrina*, *Faidherbia*, *Gliricidia*, *Inga*, *Leucaena*, *Parkinsonia*, *Pilhecel-lobiuin*, *Prosopis*, *Robinia*, and *Sesbania*. Thus, intercropping under or between fast-growing trees, in a fallow phase, is one of the approaches while finding alternative to shifting cultivation. A farming-system approach based on the watershed management has been advocated as an alternative to shifting cultivation.

20.5.2 TAUNGYA SYSTEM

The Taungya system is like an organized and systematically managed shifting cultivation. The word is reported to have originated in Myanmar (Burma) and tauang means hill, ya means cultivation, that is, hill cultivation. It involves cultivation of crops in forests or forest trees in crop-fields and was introduced to Chittagong and Bengal areas in colonial India in 1890. Later it had spread throughout Asia, Africa, and Latin. The taungya (taung = hill, ya = cultivation) is a Burmese word coined in Burma in the 1850s. The taungya system was introduced into India by Brandis in 1890 and the first taungya plantations were raised in 1896 in North Pradesh and the north-eastern hill region. In southern India, the system is called "kumari." It is practiced in areas with an assured annual rainfall of over 1200–1500 mm.

20.5.3 HOME GARDENS/HOMESTEADS CULTIVATION

A homestead is an operational farm unit in which a number of crops including tree crops are grown with livestock, poultry, and/or fish production, mainly for the purpose of meeting the routine basic needs of the farmer. It is an age-old practice in coastal states particularly in Kerala, Tripura, Assam, north-eastern states, and parts of West Bengal and Andaman and

Nicobar Islands. Although the home gardens appear to be a mixture of trees, shrubs, and herbs, a certain general pattern seems to exist. There is wide variation in the intensity of trees, species, and crops based on the size of holding, needs of the people residing in homesteads, and microclimate. Coconut, areca nut, guava, mango, citrus, tamarind, jackfruit, papaya, banana, moringa, sesbania, custard apple, and many multipurpose trees are the major trees found grown in home gardens.

Domestic animals and poultry are the main components of homesteads; therefore, sometimes forages like stylo (*Stylosanthes guianensis*/*Stylosanthes hamata*), guinea grass, Guatemala, Napier and others are also grown frequently. The components are so intimately mixed in horizontal and vertical strata as well as in time that a complex interaction exists among soil, plants, other aboveground and belowground components, nutrition, and environmental factors. There is critical competition both for light and nutrition. The holder chooses his crops and crop combinations based on his wisdom, needs, and perceptions acquired over generations of experience.

Many multipurpose trees having productive and protective functions are integrated into the system in different spatial and temporal arrangements. Most of these trees closely interact with agricultural crops when grown. These trees include teak (*Tectona grandis*), jack trees (*Artocarpus* spp.), *Casuarina equisetifolia*, *Mangifera india*, *Ceiba pentandra*, *Leucaena leucocephala*, *Grevillea robusta*, *Bambusa arundinacea*, *Erythrina variegata*, and *Gliricidia sepium* (both good support to black pepper). *Thespesia populnea* is common in low lying homesteads and the wood is commonly used for agricultural implements. Mangroves form an essential part of homesteads of backwater areas in lowlands. These are commonly used as fuelwood. Coconut and pandanus (*Pandanus tectorius*) can be commonly seen near canals and backwaters. Palmyrah palm (*Borassus flabellifer*) is common multipurpose palm in coastal Andhra Pradesh. Fish and shrimp culture in backwater channels and in association with mangroves is the main activity in homestead. Thus, multitude of crop species in the homesteads helps to satisfy primary needs of the farmer such as food, fuel, fodder, timber, and cash. This in spite of high intensity of cropping also helps to conserve fertility by nutrient cycling or organic manuring or mulching and increased microbial activity in the rhizosphere of crops. Among social benefits are—high family labor utilization and risk minimization. This system also helps in checking soil erosion (due to high intensity of vegetation cover), environmental health, and conserving biodiversity.

20.5.4 PLANTATION-BASED AGROFORESTRY SYSTEMS

Modern commercial plantation crops such as rubber, coffee, poplar, eucalypts, and oil palm represent a well-managed and profitable stable land-use activity in tropics. The scope for integrative practices involving plant associations in these commercial plantations is limited, except during the early phases of plantation when some intercropping is feasible; the commercial production of these crops is aimed at having a single commodity. Some of the plantation crops such as coconut and palms have been cultivated since very early times but their economic yield remained low. However, the research attention and commercial yields of these crops have increased substantially.

Contrary to common belief, a substantial proportion of tropical plantation crops is grown by small farmers. Most of the coconut production in India, and other countries comes from small holdings, in which coconut-palm is integrated with a large number of annual and perennial crops such as clove, cardamom, coffee, cacao, cassava, yams, fodder grasses, and legumes. Grazing under coconut and cashew nut is also common. In India, small holders grow cashew trees (in wider space) with other crops.

In the last two decades, most of the data have come from coconut-based systems in India, related to intercropping under coconuts, integrated mixed farming in small holdings, grazing under coconuts, factors favoring intensification of land use with coconuts, and multistory tree gardens. Important cereals grown with coconut include rice, finger millet, and maize; pulses such as pigeon pea, green gram, black gram, cowpea, soybean, oilseed crop such as groundnut; root crops such as sweet potato, yams, elephant foot-yam and taro; spices and condiments such as ginger, turmeric, cinnamon, clove, chilies, and black pepper; fruits such as pineapple, mango, banana, and papaya; cash crops such as cotton and sugarcane; and plantation crops such as areca nut, cacao and coffee. Many trees such as species of Erythrina, Ceiba, and Cordia find a place in the coconut-based system.

20.5.5 SCATTERED TREES ON FARM LANDS

The practice of growing agricultural crops under scattered trees on farm lands is old and does not seem to have changed for centuries. Though worldwide list of such trees is long, some of them have received more attention than others, for example, *Prosopis cineraria* in north-western India and Poplars in north India. The species diversity in these systems is very much related to ecology. With increase in rainfall, the species diversity and system complexity increases. Thus, there is a proliferation of more diverse multistoried home gardens in humid areas and less diverse, two-tiered canopy of configurations (trees + crop) in drier areas.

Trees are grown scattered in agricultural fields for many uses such as shade, fodder, fuel wood, fruit, vegetables, and medicinal uses. Some of the practices are very extensive and highly developed in India. For example, growing *Prosopis cineraria* and Zizuphus in arid areas; *Acacia nilotica* in Indo-Gangetic plains; Grevia optiva and other tree species in hills of Uttarakhand and Himachal Pradesh; *Eucalyptus globulus* in southern hills of Tamil Nadu and *Borassus flabellifer* in peninsular coastal regions. There are strong convictions for the acceptance of these trees on agricultural fields since time immemorial.

Farmers retain trees of *Acacia nilotica*, *Acacia catechu*, *Dalbergia sissoo*, *Mangifera indica*, *Ziziphus mauritiana*, and *Gmelina arborea* are preferred in Gujarat with crops. Farmers in subhumid terai region of Uttarakhand and Uttar Pradesh prefer *Dalbergia sisoo*, *Syzygium cumnii*, and *Trewia nudiflora*. In Bihar, *Dalbergia sissoo*, *Litchi chinensis*, and mango are frequently grown on fields. Every part of the palm is used by common man: the leaves for thatching, trunk as pillar or timber, fruit is roasted and consumed, the radicle of germinating seeds is roasted, and beverage (alcohol) is extracted from spadix, which is also used to prepare jaggery and vinegar. Other most common trees found on farmers' fields are *Azadirachta indica*, *Moringa oleifera*, *Tamarindus indica*, *Ceiba pentandra*, *Anacardium occidental*, and *Cocos nucifera* and fruits such as banana, custard apple, guava, and pomegranate.

20.5.6 TREES ON FARM BOUNDARIES

Trees grown in agricultural fields are also often and usually grown on farm boundaries. In northern parts of India, particularly in Haryana and Punjab *Eucalypts* and *Populus* are commonly grown along the field boundaries or bunds of paddy fields; other trees which are found grown as boundary plantations or live hedge include: *Dalbergia sissoo* and *Prosopis juliflora*. Farmerrs of Sikkim, grow bamboo (Dendrocalamus, Bambusa) all along the irrigation channels. In coastal areas of Andhra

Pradesh, *Borassus* is the most frequent palm. In Andaman, farmers grow *Gliricidia sepium*, *Jatropha* spp, *Ficus*, *Ceiba pentandra*, *Vitex trifolia*, and *Erythrina variegata* as live hedges. At many places succulents like *Agave* and many cactoids are grown as common live fence. Many of the boundary plantations also help as shelterbelts and windbreaks, particularly in fruit orchards. In Bihar, *Dalbergia sissoo* and *Wendlandia exserta* are most common boundary plantations.

20.5.7 WOOD LOTS

In many parts, farmers grow trees in separate blocks as woodlots along with agricultural fields. Now the practice is expanding fast due to shortage of fuelwood and demand of poles or pulpwood in industry. For example, bamboo poles are in great demand for orange orchards in Nagpur area and *Eucalyptus* and Poplars for match industries. These days woodlots are being raised mostly on large farms due to increase of labor costs and labor management, lack of irrigation, and risk of crop investments. Woodlots of casuarina, bamboo, poplar, *Eucalyptus*, *Leucaena leucocephala*, red sandal (*Pterocarpus santalinus*), and *Dalbergia sissoo* have become popular in many parts of the country.

20.5.8 SYSTEMS FOR SOIL CONSERVATION OR AMELIORATION

About 150 million hectares of land in India is subject to serious wind and water erosion, of which 69 million hectares are critically affected. About 4 million hectares is suffering from degradation due to ravines and gullies 11.3 million hectares as riverian land. Coastal sandy areas and steeply sloping lands and more than 9 million hectares are salt affected. The deep and narrow gullies are best controlled by putting them to permanent vegetation after closure to grazing. Afforestation with suitable tree species such as *Acacia nilotica*, *Azadirachta indica*, *Butea moonosperma*, *Prosopis juliflora*, *Dalbergia sissoo*, *Tectona grandis*, *Bambusa* spp. and *Dendrocalamus*, and other adaptable species such as grasses like *Dichanthium annulatum*, *Bothriochloa pertusa*, *Cynodon dactylon*, and *Sehima nervosum* will help in stabilizing ravines and gullies and checking their spread. In recent times, due to increase in population pressure these woodlots have shrunk at a fast rate.

20.5.9 SHELTER BELT

Arid regions witness very high wind velocity throughout the year. Farmers build shelterbelt (kana bundi) by either small dead wood or local vegetation to check wind velocity within safer limits. *Crotalaria burhia*, *Leptadenia pyrotechnica* and *Aerva pseudo tomentosa* bushes are planted in 20–25 m apart rows across the wind direction. Between the lines of these shrubs grasses such as *Cenchrus ciliaris*, *C. setigerus*, *Lasiurus sindicus*, etc. are planted on leeward side of each break. This permanent vegetation helps accumulate sand near them which is again spread in the field. This also helps increase crop yields along the lines.

20.5.10 TREES ON RANGELANDS

As pointed out above, *Salvadora oleoides*, *Capparis decidua*, *Acacia nilotica*, *A. leucophloea*, *P. cineraria*, and now *Prosopis juliflora* also are most frequent trees on common community grazing lands. In tropics coconut is most common tree on pasture lands. Cattle raising usually involves grazing on these pastures. In some cases special fodder plants including legumes are also cultivated.

20.6 AGROFORESTRY IN RECENT PAST

Especially during the last three to five centuries spectacular improvements have been made in tree plantation. The most important agroforestry practice is known from the Kangeyam tract of Tamil Nadu, where *Acacia leucophloea* + *Cenchrus setigerus* in silvi-pasture system was perfected. Similarly, in ravines of Yamuna and Chambal, trees, shrubs, and bamboos with grasses were planted for rearing milk producing Jamunapari breed of goats and sheep. Scattered trees with khejri or mehndi in association with bajra, jowar, and chilies were grown in the semiarid area of Tamil Nadu. Tree plantation continued as demarcation and control against wind erosion throughout the country. Plantation of khejri trees for various uses on on-farm system was a common practice in Rajasthan.

In coastal areas, *Casuarina equisetifolia* and other trees were grown in association with crops on farm lands for cash and to generate small timber. Live hedges were common as an agroforestry practice in which mehndi, *Agave sisalana*, and *Euphorbia* species were common. In paddy-growing areas *Pongamia glabra* and *Sesbania grandiflora* were grown, lopped annually and their leaves applied to fields as green manure. Application of green manure to paddy field was common in Madhya Pradesh and Uttar Pradesh. In Western Ghats Terminalia leaves were harvested, spread on land, burnt and then paddy, ragi, and millets were sown Multistory homesteads or home gardens were in existence in Kerala, Karnataka, Tamil Nadu, Tripura, Assam, and other northeastern states as an important agroforestry practice. This practice is still followed in these states.

20.7 SIGNIFICANCE OF FEED AND FODDER

Livestock rearing in India is changing fast and there has been a rise in demand of milch cattle as compared to dual or draught breeds. Population of indigenous breeds like Haryana, Nagori, Khilar, that is, dual and draught purpose breeds has declined more than milch breeds. In this age of market economy, the agri-economy and milk production has to compete for growing fodder on good quality land, required for high productivity and reproductive efficiency of dairy animals. Hence, its significance can be understood from the following points.

1. *Economy in production*: Feed and fodder cost constitute about 60–70% of cost of milk production thus cultivated fodder has an important role in meeting requirement of various nutrients and roughage in our country to produce milk most economically as compared to concentrates. Feeding not only meets nutrient requirement but also fills the rumen to satisfy the animals. Feed has to meet requirement of cattle maintenance, production, and requirement of microbes to promote digestion.
2. *Better feeding for ruminants*: In view of the peculiar digestive system, provided by nature, ruminants need feeds, which not only meet their nutritional requirements but also fill the rumen and satisfy the animal. In view of microbial digestion system the

feeds have to meet requirements of the animal, its production as well as the needs of microbes for promoting digestion. The fodder crops meet these requirements very effectively and hence are important for ruminant production system. As evident from reports that mixed with coarse roughages, like wheat straw, its intake and digestion are improved.

3. *Good source of critical elements*: Fodder from common cereal crops such as maize, Sorghum, and Oats are rich in energy and the leguminous crops such as lucerne, berseem, and cowpea are rich in proteins. These leguminous crops are good source of major and micro minerals, which are critical for rumen microbes as well as animal system. The green fodder crops are known to be cheaper source of nutrients as compared to concentrates and hence useful in bringing down the cost of feeding and reduce the need for purchase of feeds/concentrates from the market. In case surplus fodder is available in some season it can be stored in form of silage or hay for lean season.

20.8 FORAGE VALUE

The forage value of any feed depends on the combination of its palatability, nutritive value, and digestibility. These need to be sufficiently high for an animal to take-in its daily requirement of energy, protein, and minerals. The minimum requirement varies with the type of animal, the desired result, and climatic conditions. For 50-kg wether, for example, the daily maintenance requirement is eight mega joules of metabolizable energy, 95 g of crude protein, 5 g of calcium, 1 g of magnesium, and 2 g of sodium. To acquire this amount of energy means taking in 500 g of digestible organic matter, or 1.1 kg dry weight of feed that is 55% digestible.

The intake of sufficient energy and nutrients by an animal cannot be predicted from separate analysis of a plant's nutrient content, digestibility or palatability. Despite this, the wider use of trees and shrubs as forage is often advocated on the basis of such analyses. While these can serve as some guide to the value of species, they must be regarded with caution for several reasons. Firstly, chemical analysis commonly overestimates digestibility, particularly of protein, as it does not take into account the fact that protein is often bound to lignin and tannins which can prevent its breakdown in animals. Secondly, digestibility can be a poor indicator of forage value as shown in a study of seven common Australian browse plants where there was no relationship between true digestibility and the amount voluntarily eaten by sheep. Thirdly, palatability can vary seasonally and between animals and cannot therefore be assessed on the basis of the occasional consumption of browse.

20.9 FORAGE CONTRIBUTION FROM THE TREE SPECIES

Trees and shrubs provide valuable fodder to domestic herbivores (Panday, 1982; Robinson, 1985). At least 75% of the shrubs and trees of Africa serve as browse plants, while in South America and Australia, natural stands of trees such as *Prosopis* and *Acacia* sustain the pastoral industries of arid regions in drought.

Genuinely, productive agroforestry systems with animals involve tree legumes which provide high yields of quality forage. The principal species used for this purpose are shown in Table I . Species are available to suit a wide range of environmental and management requirements; however, the quality of forage tree legumes varies greatly from very high (*Leucaena leucocephala*) to quite low (*Acacia aneura*) (Norton, 1994).

The leaves, flowers, pods, and fine stems from the edible component of tree and shrub species and the nutritive value of this material is related to intake, digestibility, chemical composition, and the presence or absence of anti-nutritive factors. The pods in particular are a valuable high protein forage source for livestock, especially in dry periods. Felker and Bandurski (1979) suggested that species such as *Prosopis* sp. *Gleditsia triacanthos*, and *Faidherbia albida* could produce from 3–10 t pods/ha depending on the ecological zone and tree density.

Under natural conditions, a large proportion of the foliage of tree species is out of reach of the grazing animal, so that cutting or lopping may be necessary. Natural leaf fall through senescence also can be an important component of the diet of some grazing animals. In Africa, goats thrive on the leaf fall of *Acacia meliflora* (Dougall and Bogdan, 1958). Lowry (1989) reported that the leaf and pod fall from *Albizia lebbeck* in naturalized stands in northern Australia could be an important supplement to grazing cattle.

The dry matter digestibility (in sacco and in vitro) of tree legume leaf varied from high (>60% for *Sesbania sesban*, *Leucaena leucocephala*, and *Gliricidia sepium*) through moderate (50–60% for *Albizia lebbeck* and *Codariocalyx gyroides*) to low (30–50% for *Albizia chinensis*, *Acacia aneura*, and *Calliandra calothyrsus*) (Bamualim, 1981; Robertson, 1988; Vercoe, 1987). Secondary plant compounds such as tannins are present in the foliage of many tree species (e.g., *Calliandra calothrysus*, *Leucaena diversifolia*, and *Acacia aneura*) and may reduce the digestibility of protein and limit the intake of the forage by ruminants (Norton, 1994).

The nutritional quality of *Leucaena leucocephala* is regarded as excellent (Jones, 1979). This can be related to a number of factors including excellent palatability and digestibility, balanced chemical composition of minerals and protein amino acids, low fiber content, moderate tannin content to promote bypass protein value, and a biological solution to the problem of toxicity of the nonprotein amino acid mimosine and its degradation products 3, 4- and 2, 3-hydroxypyridine (Jones and Lowry, 1984).

20.10 BIODIVERSITY IN FORAGE RESOURCES

Indian subcontinent is one of the world's mega centers of crop origin and crop plant diversity, as it presents a wide spectrum of ecoclimate ranging from humid tropical to semiarid and temperate to alpine. The Indian gene centre possesses a rich genetic diversity in native grasses and legumes. There are reports of 245 genera and 1256 species of Gramineae of which about 21 genera and 139 species are endemic. One-third of Indian grasses are considered to have fodder value. Most of the grasses belong to the tribes Andropogoneae (30%), Paniceae (15%), and Eragrosteae (9%). Similarly, out of about 400 species of 60 genera of Leguminosae, 21 genera are reported to be useful as forage. The main centers of genetic diversity are peninsular India (for tropical types) and northeastern region (for subtropical types) besides some microcenters for certain species.

Major forage genera exhibiting forage biodiversity include legumes such as *Desmodium*, *Lablab*, *Stylosanthes*, *Vigna*, *Macroptelium*, *Centrosema*, etc.; grasses such as *Bothriochloa*, *Dichanthium*, *Cynodon*, *Panicum*, *Pennisetum*, *Cenchrus*, *Lasiurus*, etc. and browse plants such as *Leucaena*, *Sesbania*, *Albizia*, *Bauhinia*, *Cassia*, *Grewia*, etc. These genera besides many others form an integral part of feed and fodder resources of the country. The country is further endowed with the rich heritage of traditional know-how of raising, maintaining and utilizing forage, feed, and livestock resources.

20.11 CONCLUSION

Agroforestry for fodder and forage crops technologies will play a major role in the synthesis of sustainable farming systems for economic prosperity of farmers. It may take some more time to make it popular through development agencies. It is essential to moderate the expectations about the outcome and output level of agroforestry, and its merit of service functions may economically be valued in time to come.

Importance of fodder and forage production in maintaining food security as well as nutritional security has been felt since long. The overall scene of fodder and forage production is very alarming and corrective measures have to be taken to improve this problem. A comprehensive grazing policy needs to be formulated and both grazing and fodder and forage cultivation has to be considered complementary to each other and simultaneous efforts are required to improve both. Fodder tree improvement programs for higher leaf fodder have to be initiated. For the improvement of grasslands, its management needs to be considered holistically promoting interaction between grassland, livestock, and grazing communities. Therefore, the vast natural resource can serve human society substantially, more particularly grazing communities. A favorable policy environment in terms of access to microcredit and assured market will have to be provided and simultaneously there is need to address the socioeconomic and technical constraints.

The degree of success will continue to vary due to its location specificity, appropriate choices, and a number of local factors. Much of the conventional wisdom has been challenged by recent research and farming experience in the context of adoption of a perfected technology. There is a serious revision on many fronts. It is not certain that technique or technology perfected today will stand the test of experience tomorrow. Hence, a continuum of interactions of research efforts and farm experience should flow to realize the potential of agroforestry for fodder and forage. Overall, only when the fact that a tree planted today is really for use in the next generations is accepted, agroforestry will work.

KEYWORDS

- **agroforestry**
- **systems**
- **biodiversity**
- **forage value**
- **fodder**

REFERENCES

Bamualim, A. Nutritive Value of Some Tropical Browse Species in the Wet and Dry Tropics. M.Sc. Thesis, Lames Cook University of North Queensland, 1981.

Dougall, H. W.; Bodgan, A. V. Browse Plants of Kenya with Special Reference to Those Occurring in South Baringo. *East Afr. Agric. J.* **1958,** 236–245.

FAO. *The State of Food and Agriculture: Livestock in the Balance*; Food and Agriculture Organization of the United Nations: Rome, Italy, 2009.

Felker, P.; Bandurski, R. S. Uses and Potential Uses of Leguminous Trees for Minimal Energy Input Agriculture. *Econ. Bot.* **1979,** *33*, 172–184.

Jones, R. J. The Value of *Leucaena leucocephala* as a Feed for Ruminants in the Tropics. *Wld. Anim. Rev.* **1979,** *31*, 13–23.

Jones, R. J.; Lowry, J. B. Australian Goats Detoxify the Goitrogen 3-Hydroxy-4(IH) Pyridone After Rumen Infusion from Indonesian Goats. *Experimentia* **1984,** *40*, 1435–1436.

Lowry, J. B. Agronomy and Forage Utilization of *Albizia lebbeck* in the Semi-arid Tropics. *Trop. Grassl.* **1989,** *23*, 84–91.

MacOpiyo, L.; Angerer, J.; Dyke, P.; Kaitho, R. In *Experiences on Mitigation or Adaptation Needs in Ethiopia and East African Rangelands*. Livestock and Global Climate Change. Proceedings, of the International Conference in Hammamet, May 17–20, 2008; Rowlinson, P, Steele, M, Nefzaoui, A. Eds.; British Society of Animal Science, Cambridge University Press: Cambridge, UK, 2008; pp 64–67.

Neufeldt, H.; Jahn, M.; Campbell, B. M.; Beddington, J. R.; DeClerck, F.; De Pinto, A.; Gulledge, J.; Hellin, J.; Herrero, M.; Jarvis, A. Beyond Climate-Smart Agriculture: Toward Safe Operating Spaces for Global Food Systems. *Agric. Food Secur.* **2013,** *2*, 12.

Ngigi, M. *The Case of Smallholder Dairying in Eastern Africa*; Environment and Production Technology Division Discussion Paper No. 131. The International Food Policy Research Institute: Washington, DC, USA, 2005.

Norton, B. W. The Nutritive Value of Tree Legumes. In *Forage Tree Legumes in Tropical Agriculture*; Gutteridge, R. C., Shelton, H. M., Eds.; CAB International: Wallingford, UK, 1994; pp 177–191.

Panday, K. K. *Fodder Trees and Tree Fodder in Nepal*; Swiss Development Corporation: Berne, Switzerland, 1982; p 107.

Robinson, P. J. Trees as Fodder Crops. In *Attributes of Trees as Crop Plants*; Cannell, M. G. R, Jackson, J. E., Eds.; Institute of Terrestrial Ecology: Huntingdon, UK, 1985; pp 281–300.

Robertson, B. M. The Nutritive Value of Five Browse Legumes Fed as Supplements to Goats Offered a Basal Rice Straw Diet. M.Agr. Studies Thesis, The University of Queensland, 1988.

Thorlakson, T.; Neufeldt, H. Reducing Subsistence Farmers' Vulnerability to Climate Change: Evaluating the Potential Contributions of Agroforestry in Western Kenya. *Agric. Food Secur.* **2012,** *1*, 15.

Thornton, P.; Herrero, M. In *Priority Livestock Development Issues Linked to Climate Change*. Livestock and Global Climate Change. Proceedings, International Conference in Hammamet, May 17–20, 2008. Rowlinson, P., Steele, M., Nefzaoui, A., Eds. British Society of Animal Science, Cambridge University Press: Cambridge, UK, 2008; pp 21–24.

Vercoe, T. K. In *Fodder Potential of Selected Australian Tree Species*. Australian Acacias in Developing Countries. ACIAR Proceedings, 16, Canberra; Turnbull, J. W., Ed.; 1987; pp 95–100.

World Bank. Global Economic Prospects January 2014. The World Bank: Washington, DC, USA, 2014.

CHAPTER 21

PRESERVATION OF FORAGE CROPS

PAMPI PAUL*, MAHESH B. TENGLI, and B. S. MEENA

Division of Dairy Extension, National Dairy Research Institute, Karnal 132001, Haryana, India

Corresponding author. E-mail: pampindri@gmail.com

ABSTRACT

Storing forage for use in the nongrowing season is an ancient practice which probably originated in countries where climatic conditions for preservation of food were favorable and livestock were important. Storage of straws and crop residues for feed is probably as old as crop production. In the Indian context, the feed resources for animals include mainly native plants, fodder trees, and farm by-products. On the contrary, the use of forage crops like pasture plants are limited to the advanced regions where mainly dairy farming is practiced. It is important to make good use of the feed resources which are produced in these regions and are inexpensively and easily available. It is necessary to solve the production and utilization problems of forage crops, like pasture plants, by considering the land use, cost of production and feed needs in the dairy farming, etc. As to the use of farm and food processing by-products, the condition of the production, availability, prices, and methods for procurement like transportation should be examined. It is necessary to adjust the moisture content of the silage materials, accelerate fermentation and improve the nutritional value of the feed by adding these by-products to the materials of feed crops or native grasses at the time of silage making.

21.1 INTRODUCTION

Livestock has an imperative role to play in farmers' life and livelihood as well as in the Indian economy as it is rural and agrarian oriented. Since the long back, animals have played an important role in the existence and development of human economy. They serve as a source of family savings and economic security in the event of failure of food crops and natural calamities. They are a vital factor of human culture such as in religious activities and ceremony, exhibition and shows, fighting, racing, pets and recreation, and status symbol. Dairying in particular is increasingly popular because it means cattle, which represents milk, wealth, draught power and manure can be used to bring in a cash income as well, without having to sell them. However, because of the erratic and poor rains, and the long dry season, dairying is only viable if the cattle are supplemented with feed, otherwise the milking cows dry off early in the year and there is no milk to sell until they calve again. Also, with poor feeding, the cows often produce calves once every 2–3 years. A dairy cow needs to produce a calf every year so that she will produce milk every year. However, dairy farmers cannot afford to feed expensive commercial dairy concentrates as a supplement. For these grazing animals need forages as basal or main food, therefore, most important aspect of the husbandry of grazing animals is that can obtain milk productivity and can give it return back to its owners.

21.2 IMPORTANCE OF PRESERVATION OF FORAGES

Storing forage for use in the nongrowing season is an ancient practice which probably originated in countries where climatic conditions for preservation of food were favorable and livestock were important. Storage of straws and crop residues for feed is probably as old as crop production. In the Indian context, the feed resources for animals (cattle and other ruminants) include mainly native plants, fodder trees, farm by-products (leaves and stems except for food use, straws, grains, brans, etc.), and food processing by-products (cassava meal, soybean curd, molasses, etc.). On the contrary, the use of forage crops like pasture plants are limited to the advanced regions where mainly dairy farming is practiced. It is important to make good use of the feed resources which are produced in these regions and are inexpensively and easily available. It is necessary to solve the production and utilization problems of forage crops, like pasture plants, by considering the land use, cost of production and feed needs in the dairy farming, etc. As to the use of farm and food processing by-products, the condition of the production (time, form, moisture content, amount, etc.), availability, prices, and methods for procurement like transportation should be examined. It is necessary to adjust the moisture content of the silage (total mixed ration, TMR) materials, accelerate fermentation, and improve the nutritional value of the feed by adding these by-products to the materials of feed crops or native grasses at the time of silage making.

In many regions, the climate conditions are characterized by the rainy season and the dry season. In the rainy season, as the growth of herbage including native grasses and pasture plants is vigorous, sufficient feed becomes available and in most cases animals can be fed enough by cut-and-carry or grazing However, in the dry season, not enough feed tends to be secured and fed to animals, because herbage plants stop growing or dying due to the continued dry condition during the dry season. Thus, animals suffer malnutrition, reduction of milk production, loss of body weight, outbreak of diseases, and reproduction disorders. These marked reductions in animal productivity cause mainly the stagnation of income growth in animal farming and become a limitation factor for increasing the animal keeping number. In the advanced regions where dairy farming is practiced, pasture plants, etc. suitable for the tropical regions are introduced and selected and are widely spreading. Therefore, it is vital to produce good quality roughage for the dairy animals year-round by increasing the cropping area.

21.3 PRESERVATION OF FORAGE

Usually forage is preserved as either hay or silage. Whether to preserve as hay or silage it depends on different factors. In hay production, the crop is dried so that it is essentially biologically inactive both with respect to plant enzyme activity and microbial spoilage. The low moisture content also permits easier transportation by reducing the weight per unit of dry matter (DM). Hay making is dominant in those areas of the world where good drying conditions prevail. However, it may also be used in humid climates where ensiling has been considered too difficult because of forage characteristics, high temperatures, or tradition.

21.4 HOW FORAGES SHOULD BE CONSERVED FOR THE DRY SEASON: HAY OR SILAGE?

Forages can be conserved as hay or as silage. Natural pasture and planted pastures can be made into hay, provided they are cut early enough—by March—to conserve the nutrients, especially protein, before they decline in the plant. However, even in March, it is often too wet to dry the pasture successfully and special machinery, which is very expensive, has to be used to assist the forage to dry quickly. Other forage crops such as maize, forage sorghum pennisetums are too thick-stemmed to dry successfully as hay. It is difficult to make hay successfully from legumes as they drop their leaves very readily, either when handing the dried material or if allowed to get too mature before cutting. Silage is considered the better way to conserve forage crops. A forage crop can be cut early and only has to have 30% dry matter to be ensiled successfully. There is no need to try and dry out the plant material any more than that, so wet weather is not such a constraint as it is with making hay. This means the crop can be cut in March, depending on when it was planted, as if legumes are intercropped with the forage, they will not lose their leaves at cutting. So based on such factors, we have decided that for preservation of forages which one will be good, whether hay or silage, based on the such suitability of the farmers. Here, we will discuss about silage and hay in details.

21.5 WHAT IS SILAGE?

Green fodder is considered an economical source of nutrients for the dairy animals. While increase in green fodder production per hectare of land has been emphasized, it is equally important to conserve green fodder to ensure regular supply for feeding especially during the lean periods. Conserving green fodder in the form of silage is one of best options available to ensure regular supply of quality fodder through different seasons of the year.

Forage which has been grown while still green and nutritious can be conserved through a natural "pickling" process. Lactic acid is produced when the sugars in the forage plants are fermented by bacteria in a sealed container ("silo") with no air. Forage conserved this way is known as "ensiled forage" or "silage" and can be kept for up to 3 years without deteriorating.

21.6 PRINCIPLES OF SILAGE MAKING

After harvesting of crops they continue to respire as long as they remain adequately hydrated and oxygen is available. The oxygen is necessary for the physiological process of respiration, which provides energy for functioning cells. In this process, carbohydrates (plant sugars) are consumed (oxidized) by plant cells in the presence of oxygen to yield carbon dioxide, water and heat: sugar + oxygen → carbon dioxide + water + heat.

Once in the silo, certain yeasts, molds, and bacteria that occur naturally on forage plants can also reach populations large enough to be significant sources of respiration. In the silage mass, the heat generated during respiration is not readily dissipated, and therefore the temperature of the silage rises. Although a slight rise in temperature from 80°F to 90°F is acceptable, the goal is to limit respiration by eliminating air (oxygen) trapped in the forage mass. Some air will be incorporated into any silo during the filling process, and a slight increase in silage temperature is likely. These temperature increases can clearly be limited by harvesting at the proper moisture content and by increasing the bulk density of the silage. Generally, it is desirable to limit respiration during the fermentation process by using common sense techniques that include close inspection of the silo walls prior to filling, harvesting the forage at the proper moisture content, adjusting the chopper properly (fineness of chop), rapid filling, thorough packing, prompt sealing, and close inspection of plastics for holes.

21.7 CHARACTERISTICS OF GOOD QUALITY SILAGE

i. Bright, light green-yellow or green-brown in color.
ii. Lactic acid odor with no butyric acid and ammonia odor.
iii. Firm texture with softer material.
iv. Moisture should be in range of 65–70%.
v. Lactic acid 3–14%.
vi. Butyric acid less than 0.2%.
vii. pH range is in between 4.0 and 4.2.

21.8 ADVANTAGES OF SILAGE

- Stabile composition of the feed (silage) for a longer period (up to 5 years).
- Plants can be harvested at optimal phase of development and are efficiently used by livestock.
- Reduction of nutrient losses which in standard hay production may amount to 30% of the dry matter (in silage is usually below 10%).
- More economical use of plants with high yield of green mass.
- Better use of the land with two to three crops annually.
- Silage is produced in both cold and cloudy weather.
- The fermentation in silage reduces harmful nitrates accumulated in plants during droughts and in overfertilized crops.
- Allows by-products (from sugar beet processing, maize straw, etc.) to be optimally used;
- Requires 10 times less storage space compared to hay;
- Maize silage has 30–50% higher nutritive value compared to maize grain and maize straw;
- 2 kg of silage (70% moisture) has the equal nutritive value of 1 kg of hay.

21.9 DISADVANTAGES OF SILAGE

- Silage is not interesting for marketing as its value is difficult to be determined.
- It does not allow longer transportation.
- The weight increases manipulation costs.
- Has considerably lower vitamin D content compared to hay.

21.10 FERMENTATION PROCESS

Silage, which is succulent roughage, is made by keeping chopped silage materials air-tight in a suitable container (silo) to undergo mainly lactic acid fermentation with the aim of storing feed. The principle of silage is the same as that in making pickles.

The fermentation process of silage is as follows:

1. The first stage

- The packed raw materials are still respiring immediately after chopping, and consume oxygen.
- The temperature will rise to about 32°C around 4 days after packing.

2. The second stage

- Acetic acid production begins by fermentation with acetic acid bacteria during the respiration in the first stage.
- The silage pH slowly changes from about 6.0–4.0.

3. The third stage

- Lactic acid fermentation begins by lactic acid bacteria about 3 days after packing chopped materials.
- Acetic acid fermentation by acetic acid bacteria decreases, and then acetic acid production declines.

4. The fourth stage

- Lactic acid production continues for about 2 weeks.
- The temperature goes down slowly to about the normal atmospheric temperature.
- The pH decreases to about 4.0, and the activity of the various bacteria ceases.

5. The fifth stage

- If the reaction proceeds smoothly up to the fourth stage, it enters a stable phase with a low pH condition, and high-quality silage is made.
- The lactic acid fermentation completes in about 20 days, and the silage product is finished.
- If the lactic acid production is insufficient, butyric acid fermentation begins and quality deterioration occurs.

21.11 CRITICAL FACTORS AFFECTING PRODUCTION OF GOOD QUALITY SILAGE

1. Type of silo—surface silo is best due to ease of ensiling
2. Dry matter of fodder—ideal is 30–35%
3. Chop length of fodder—ideal is 2–3 cm, easy to get compacted
4. Pressing/compaction of fodder—as quick as possible to minimize aerobic fermentation
5. Sealing of silo—to check inflow of air and water into silo

21.12 THE ENSILING PROCESS

Silage fermentation can be classified as either primary (desirable) or secondary (undesirable) (Pahlow et al., 2003). Primary fermentation is carried out by lactic acid producing bacteria and is classified as homofermentative (the one product of fermentation is lactic acid) and heterofermentative (multiple products of fermentation are lactic and acetic acids and ethanol). Secondary fermentation is carried out mainly by enterobacteria (which produce lactic, acetic, succinic, and formic acids, and ethanol), *clostridia* (produce butyric acid), and yeasts (produce ethanol). Lactic acid production is preferred over the other fermentation products due to faster and lower pH drop (stronger acid), and limited silo shrink. Shrink occurs from plant and microbial respiration, fermentation, runoff, and loss of volatile organic compounds. If anaerobic and acidity conditions are not met, silage is more prone to shrinking during storage compared to hay. Good fermentation should result in DM losses of less than 10%.

21.13 PHASES OF SILAGE FERMENTATION

According to Collins and Owens (2003), overview of four phases of the silage fermentation is described as follows:

21.13.1 AEROBIC

During this period, plant cells and microbes will metabolize sugars and starch in the presence of oxygen, generating heat in the process. Silage temperature is elevated to about 90°F, and water may be lost (as seepage) because of respiration and compaction. Usually this phase of fermentation lasts for 1 day only. In this phase, it is critical to ensure good compaction, proper moisture, and good sealing, all of which lead to a rapid transition to anaerobic conditions.

21.13.2 FERMENTATION

Once anaerobic conditions are achieved, lactic acid bacteria and other anaerobes start to ferment sugars into lactic acid, mainly, and other organic acids to a lesser extent (such as acetic and propionic) that will drop the silage pH from about 6.0 to a range of 3.8–5. Alcohols such as ethanol will be generated too, but with no contribution to the acidification process. Rapid decrease in pH prevents breakdown of plant proteins and helps inhibit growth of spoilage microbes. Consequently, lactic acid production is preferred to ensure a low silo shrink. The fermentation phase usually lasts from 1 week to more than a month, depending on crop and ensiling conditions.

21.13.3 STABLE

As long as anaerobic conditions are upholding, silage can be stable for months and even up to years. However, under normal conditions, silage should be used within a year of its production. Because slow entry of air through areas that were not properly sealed can slowly deteriorate material, thus silos should be constantly checked and maintained to avoid any potential break of seal integrity.

21.13.4 FEED OUT

Once a silo is opened, it should be used as quickly as possible to avoid aerobic deterioration of the materials. When oxygen becomes available in the ensiled material, yeasts metabolize the organic acids, which in turn cause the pH to increase, and further restarts the aerobic activity (such as molds), causing greater silage spoilage. The design of a typical silo face should allow for the daily removal of approximately 6 in. of face materials. Silo opening should occur only after the fermentation phase has been completed (i.e., after 3–6 weeks).

21.14 MANAGEMENT PRACTICES FOR GOOD QUALITY SILAGE

There are several managemental factors responsible for good quality silage, some of them are:

21.14.1 CROP FACTORS

- An ideal crop to be ensiled should have adequate level of sugars to be fermented.
- Crop with low buffering capacity and standard dry matter concentration with more than 20%.
- Plants with optimum water soluble carbohydrates are required for adequate fermentation.

21.14.2 HARVEST MATURITY

Stage of harvesting of a crop determines silage quality as optimize for yield, nutritive value and fermentability of different crops also influences silage quality.

21.14.3 MOISTURE CONTENT

It affects the rate and degree of fermentation during ensiling process. Forages which contain more than 70% moisture should not be ensiled for fermentation.

21.14.4 PARTICLE SIZE

Optimum chopped grasses can lead or effect efficient silage fermentation. Once it is short, it is easier to consolidate, however too short a chop length less than 10 mm, can lead to scouring animals as it reduces the fiber available for good rumen function, even though animals may eat more. On the other hand, a long chopped silage, more than 40 mm, likely to be less effectively consolidated and leading to poor fermentation and spoilage. That is why an optimum length is required to get good quality silage. The optimum chop length for silage of greater than 20% dry matter is 10–25 mm and for silage of less than 20% dry matter is 20–40 mm.

21.15 DIFFERENT TYPES OF SILO

There are different kinds of silo to be prepared for dairy animals, based on the structures it has been classified and described as follows:

1. Stack silo

a. Silage can be made easily with a simplest silo. Plastic sheet (about 0.1 mm thick) is spread over the ground, and similarly chopped silage materials on the sheet are entirely covered with a plastic sheet. Proper tread pressure has to be applied, and complete sealing is required (Fig. 21.1).
b. Size of silo can be determined according to the number of raising animals.

FIGURE 21.1 (See color insert.) Stack silo.

2. Bunker silo

a. For this kind of silo, side walls made of wood and concrete are needed, and the interior is preferably sealed by plastic sheets. Proper tread pressure has to be applied, and complete sealing is required. Here supports are needed so that the side walls do not fall toward the outside.

b. Width of the front should be such that the total amount of silage per day can be taken out with a thickness of 20–30 cm to prevent aerobic deterioration (Fig. 21.2).

FIGURE 21.2 (See color insert.) Bunker silo.

3. Trench silo

a. A trench silo can be built by simply digging the ground, but it is better to place plastic sheets inside to prevent loss. In this case, proper tread pressure also has to be applied, and complete sealing is required. A trench silo whose interior is coated with concrete can be used for a long time.

b. Width of the front should be such that the total amount of silage per day can be taken out with a thickness of 20–30 cm to prevent aerobic deterioration.

4. Plastic bag silo

a. This is a plastic bag with the thickness of about 0.1 mm and silage materials are packed inside. In this type of silo, commercial plastic bags are used if available.
b. Plastic bags for fertilizer and feed may be reused for cost-cutting. Bags must be packed with chopped raw materials, compressed as much as possible to remove the internal air and then sealed completely (Fig. 21.3).

FIGURE 21.3 (See color insert.) Plastic bag silo.

c. The number of bags is determined freely in accordance with the operation size. Here it is required to watch out for damage on the plastic bag by field mice, birds, and dogs.

5. Tower silo

a. This kind of silo is tall cylindrical silo built above ground of masonry, wood, or enameled steel (Fig. 21.4).

FIGURE 21.4 **(See color insert.)** Tower silo.

b. Typically it is 10–90 ft (3–27 m) in diameter and 30–275 ft (10–90 m) in height with the slipform and jumpform concrete silos being the larger diameter and taller silos.
c. They can be made of many materials. Wood staves, concrete staves, cast concrete, and steel panels have all been used, and have varying cost, durability, and airtightness tradeoffs.

6. Fenced silo

a. In this type of silo, frame is made of bamboo, wooden, iron materials, etc., which are easily available locally. Shape of cross section may be circular or rectangular.

b. Inside of the silo is sealed with plastic sheets. The silos are packed with compact and chopped raw materials and must be compressed as much as possible to remove the internal air and sealed completely.

21.16 JUDGING THE QUALITY OF GOOD SILAGE

1. *Color*: In general, pale yellow indicates good quality. If the color is from dark brown to dark green, the silage underwent bad fermentation and is of bad quality.
2. *Smell*: Acidic or a sweet–sour pleasant smell indicates good quality. On the other hand, if there is a manure smell or putrid smell and it is so repugnant that one cannot put the silage near one's nose, the quality is poor.
3. *Taste*: If the silage tastes sour and there is no problem in putting it in one's mouth, the quality is good. On the other hand, if the silage tastes bitter and one cannot put it in one's mouth, the quality is poor.
4. *Touch*: When squeezing the silage tightly in a hand and then opening the hand, if the silage breaks slowly into two, that silage is of good quality. If the silage breaks into small pieces separately, the silage is deficient in moisture content. If water is dripping, the moisture content of the silage is too high.

21.17 SOME POINTS DIFFER IN HAY AND SILAGE

Hay is much suitable for marketing and transportation purpose as its weight is less than the silage and is produced under anaerobic condition. Moreover, silage is much appropriate for the places where mechanized conditions are privileged as it required infrastructural advantages. Even for both the products we need different level of moisture content such as for high-moisture silage (≤30% DM), medium-moisture silage (30–40% DM), and low-moisture (wilted) silage (40–60% DM). Low-moisture silage is referred to as haylage. When baled and wrapped, haylage is referred to as baleage. High-moisture silages are more prone to potential seepage losses (i.e., effluent or leachate from the silo), undesirable secondary fermentation (resulting in butyric acid, which results in a rancid smell), and high dry matter losses (silo shrink). On the other hand, preservation as haylage depends more on achieving adequate packing (high density) to maintain anaerobic conditions. Achieving high density at packing is more difficult in drier forage. Nevertheless, high density is one of the critical factors in haylage to maintain anaerobic conditions because microbes are less active and fermentation is lower in haylage than in higher moisture silage.

21.18 WHAT IS HAY?

Hay is defined as forage conserved under aerobic dry or limited moisture conditions. Fresh forage typically has moisture concentration between about 75% and 85% (Collins and Coblentz, 2013). Thus, the goal in hay production is to remove moisture as quickly as possible to achieve a target moisture concentration equal to or less than 20% (or a target dry matter concentration greater than 80%). This process of reducing moisture is called curing and is normally accomplished with energy provided by the sun (field curing) or by artificial barn drying using forced heated or unheated air. Moisture concentration less than 20% (preferably less than or equal to 15%) prevents plant respiration and allows for an almost complete conservation of plant nutrients for extended periods (month).

21.19 FACTORS INFLUENCE THE PROCESS OF MOISTURE LOSS FOR HAY PRODUCTION

Factors can be classified into three types: (1) forage-related, (2) weather-related, and (3) management-related (Rotz, 1995; Collins and Owens, 2003).

21.19.1 FORAGE-RELATED FACTORS

21.19.1.1 STEM THICKNESS AND WAXY CUTICLE

As stem thickness increases (solid stems), the drying process slows because of increased radial distance from the stem core to the epidermis, where water must travel to move out of the plant. It is more difficult to dry thick-stemmed, erect plants—such as corn and sorghum-type plants—as fast as plants with thinner stems, such as tall fescue, orchard grass, or Bermuda grass (Brink et al., 2014).

21.19.1.2 FORAGE SPECIES

The differences in drying rates among forage species are mostly a consequence of a high surface area to dry-weight ratio. In addition, forages with greater leaf to stem ratios dry faster because leaves dry faster than stems (Rotz, 1995).

21.19.2 WEATHER-RELATED FACTORS

The most unpredictable variable and highly correlated to deal with when making hay is the weather. In reality, very little can be done to the plant or swath to improve drying rates if the environmental conditions are not conducive to moisture loss and it is therefore difficult to isolate the effects of each factor. Favorable conditions for hay production include high temperature, high solar radiation, and moderate wind speed (up to 12 mph) in conjunction with low air relative humidity and low soil moisture. The drying rate is faster at the beginning of the drying process; however, it slows down and reaches zero when moisture equilibrium with the environment is reached.

21.19.3 MANAGEMENT-RELATED FACTORS

Hay will usually require 3–5 days or more of field curing to reduce moisture to less than 20%. It is especially important to dry hay to less than 40% moisture as soon as possible to prevent nutrient loss due to plant respiration and microbial degradation. Different management practices such as mowing, curing, window inversion, mechanical or chemical conditioning, bailing, etc. are responsible for good quality hay production.

21.20 HAY PRODUCTION PROCESS

Hay making is a way to preserve fodder and grasses. Leguminous plants, which are major source of protein, can also be conserved in this way for storage and feeding at a later stage.

21.20.1 HAY MAKING

Here, the basic principle is to reduce the moisture concentration in the green forages sufficiently as to permit their storage without spoilage or further nutrient losses. Basically, hay is dried fodder. The moisture concentration in hay must be less than 15% at storage time. Therefore, crops with thin stems and many leaves are better suited for hay making as they dry faster than these having thick and pithy stems and small leaves.

21.20.2 HARVESTING AND CURING OF HAY

If we talk about harvesting of the crops, then flower initiation stage or when crown buds start to grow is best and suitable stage to harvest. While in case of grasses preflowering or flower initiation stage is appropriate. The harvested forage should be spread in the field and raked a few times for quick drying.

21.20.3 ARTIFICIAL DRYING

Field curing is the most common during bright sunny days, which causes bleaching of the forages and loss of leaves due to shattering. Nutrients may also be leached out if the forage is exposed to rain. To avoid these losses, forages can be dried in barn by flowing hot air through the forage. The main advantages of that are nutrient losses due to leaf shattering and bleaching can be checked but sometimes artificial drying incurred cost and process become expensive.

21.20.4 TEDDING AND RAKING

Once hay curing is done in swaths or windrows, much of the solar radiation is intercepted by crop stubble or by the soil. Tedding disperses the crop so that it covers the entire field. Even tedding creates thinner crop layers that lead to more uniform moisture levels, which are useful during hay storage. Tedding can be avoided on low yielding and short crops to reduce losses associated with raking of short materials, moisture loss is usually adequate without tedding under these conditions.

Raking gathers the crop into windrows for bailing. The timing of raking with respect to crop moisture greatly affects dry matter losses associated with this step, especially leguminous hay crops.

21.20.5 BAILING

Bailing is collecting hay into packages for ease of handling, transport and storage. Different types of bales are there to use based on the situation. Collection and stacking for storage can be done mechanically or manually. The dried forage should be baled when the moisture concentration become lower than 15%. Bailing is effective as it help hays in storage and requires less space.

21.20.6 HAY STORAGE

Storage losses are about 5% for hay harvested near 15.5% moisture and stored under dry conditions. But the forage quality is decreased during storage of hay baled above 20% moisture unless steps are taken to avoid microbial growth and heating. Poor storage conditions result in much loss in dry matter and quality in high rainfall regions.

21.21 CONCLUSION

Preservation of forages is a feasible gift to dairy farmers as it is important as when the green fodder for animals are not available, preservation of forages is the only option to complement the feed basket of the dairy animals. Silage and hay making is presenting a neoclassical and green way of utilizing crop waste and convert them into excellent feedstuff for livestock, which in turn will be reflected as in milk production. At the same time, silage making is reducing dairy farm's dependency on feed stuffs that can also be used as human food. Added to this silage or hay making is opening new areas for rural entrepreneurship which can help the youth of this generation to be self-dependent.

ACKNOWLEDGMENT

The authors of this chapter highly gratitude to the editor of this book Dr. Md. Hedayetullah, assistant professor (Agronomy), Bidhan Chandra Krishi Viswavidyalaya, Kalyani, West Bengal to give us such a opportunity to write a valuable chapter which will be helpful for agricultural students as well as for research extension purpose.

KEYWORDS

- **preservation**
- **silage**
- **hay**
- **quality**

REFERENCES

Brink, G.; Matthew, F. D.; Muck, R. E. Field Drying-rate Differences among Three Cool-season Grasses. *Forage and Grazinglands* **2014,** *12* (1). DOI: 10.2134/FG-2013-0104-RS.

Collins, M.; Owens, V. N. Preservation of Forage as Hay and Silage. In *Forages: An Introduction to Grassland Agriculture*; Barnes, R. F., Nelson, C. J., Moore, K. J., Collins, M., Eds.; Blackwell Publishing: Ames, IA, 2003; pp 443–471.

Collins, M.; Coblentz, W. K. Postharvest Physiology. In *Forages: The Science of Grassland Agriculture*, 6th ed.; Barnes, R. F., et al., Eds.; Blackwell Publishing: Ames, IA, 2013; Vol. 2, pp 583–599.

Pahlow, G.; Muck, R. E.; Driehuis, F.; Oude Elferink, S. J. W. H.; Spoelstra, S. F. Microbiology of Ensiling. In *Silage Science and Technology*; Buxton, D. R., Muck, R. E., Harrison, J. H., Eds.; American Society Agronomy,-Crop Science Society of America-Soil Science Society America Publishers: Madison, WI, 2003; pp 31–94.

Rotz, C. A. Field Curing of Forages. In *Postharvest Physiology and Preservation of Forages*; CSSA Publishers: Madison, WI, 1995; pp 39–65.

CHAPTER 22

FORAGE PRODUCTION AND CLIMATE CHANGE

K. JANA[1*], A. M. PUSTE[2], J. BANERJEE[3], S. SARKAR[3], and R. J. KOIRENG[4]

[1]*AICRP on Forage Crops and Utilization, Directorate of Research, Bidhan Chandra Krishi Viswavidyalaya, Kalyani 741235, Nadia, West Bengal, India*

[2]*Department of Agronomy, Bidhan Chandra Krishi Viswavidyalaya, Mohanpur 741252, Nadia, West Bengal, India*

[3]*Department of Genetics and Plant Breeding, Bidhan Chandra Krishi Viswavidyalaya, Mohanpur 741252, Nadia, West Bengal, India*

[4]*Directorate of Research, Central Agricultural University, Imphal 795004, Manipur, India*

[*]*Corresponding author. E-mail: kjanarrs@gmail.com*

ABSTRACT

Production and productivity of crops are mainly dependent on climate. Climate is a primary determinant of agricultural productivity. Food (feed), forage (fodder), and fiber production are essential for sustaining and enhancing human welfare. Food-forage production system is most important for improving sustainability, productivity, and economic stability of rural farming community under changed climate. Hence, agriculture has been a major concern in the discussions on changed climate scenario. There is increasing evidence from climatic observations that the climate is changing.

22.1 INTRODUCTION

Production and productivity of crops are mainly dependent on climate. Climate is a primary determinant of agricultural productivity. Food (feed), forage (fodder), and fiber production are essential for sustaining and enhancing human welfare. Food-forage production system is most important for improving sustainability, productivity, and economic stability of rural farming community under changed climate. Hence, agriculture has been a major concern in the discussions on changed climate scenario. There is increasing evidence from climatic observations that the climate is changing. Intergovernmental Panel on Climate Change (IPCC, 2007) reported that the global mean temperature has risen by 0.8°C since the 1850s, with warning found in three independent temperature records over land and seas and in the ocean surface water. The Earth's average surface temperature has increased 1.3°F over the past century and is projected by IPCC to increase by an additional 3.2–7.2° over the 21st century (IPCC, 2007). Tans and Keeling (2012) reported that carbon dioxide (CO_2) levels in the atmosphere have gone up from 284 mg/kg in 1832 to 391 mg/kg in 2012 and it is mainly due to the burning of fossil fuels, with smaller contribution from land-use changes.

22.2 IMPACTS OF CHANGED CLIMATE ON FOOD-FORAGE CROPS

Agriculture is inherently sensitive to climate change and variability. Climate change and variability is mainly due to natural causes or human activities. Climate change is mainly due to emission of greenhouse gases (GHGs) and it is expected to directly impact on crop production systems for feed/food (grain), fiber, and fodder (forage). It also affects livestock health and alters the pattern and balance of trade of animal products. These impacts of climate change will vary with the degree of warming and associated changes in rainfall patterns and from one place to another. But underpinning these impacts of climate change is a number of direct effects on the physiology of fodder-forage crops grown for animal feed. Drake et al. (1997) reported that increase concentration of carbon dioxide (CO_2) in the atmosphere, which is one of the main GHGs and it enhances the production and productivity of most of the fodder-forage crops due to enhanced rates of photosynthesis (Long et al., 2006). This boost to productivity is apparent for all crops that use the C_3 photosynthetic pathway, namely, cowpea, oats, and soybean (Barbehenn, 2004). However, there are number of important forage crops, namely, maize, sorghum, sugarcane millets, etc. that have a different response to increased CO_2. They use the C_4 photosynthetic pathways. Kimball et al. (1983) reported that the leaf photosynthetic rates of C_4 plants are not substantially enhanced elevated concentrations of CO_2. Hence, yield gains in plants grown under elevated CO_2 are much more modest than for C_3 plants. Warmer temperatures affect the rate at which crops grow and develop, and potentially the survival of plants and grain at extreme of temperature. The duration from sowing to flowering and crop harvest, is determined by mean temperature and day length (Craufurd and Wheeler, 2009). As climate warms, the duration to harvest shortens, at least until a really hot optimum mean temperature is exceeded. Farmers will mostly try to adapt to these climatic changes by using new long season varieties/cultivars or different major/minor food-forage crops for their regions. Where longer season varieties cannot be used, crop yields will decline with warmer temperatures. Extremes of hot temperatures will become more frequent under climate change. Where hot days coincide with a sensitive stage of crop development, such as flowering, we find dramatic decrease in seed or grain yields because of disruption to pollination (Wheeler et al., 2000).

22.3 FORAGE DIGESTIBILITY, YIELD, AND QUALITY UNDER FUTURE CLIMATE CHANGE

Sorghum is becoming an increasing important food cum forage crop in many regions of the world. It is highly resistance to drought situation. So, it is also a suitable food cum forage crop for semiarid regions. Saeed and Nadia (1998) reported that sorghum can be a response to additional irrigation water by stem elongation and increase of yield. They were reported that the water deficit stress reduced quantitative and qualitative yield included total fresh weight, total dry weight, leaf dry weight, stem dry weight, protein yield, and leaf/stem ratio of forage millet "nutrifeed." Oats (*Avena sativa*) are also the most important feed cum forage crop during winter season. Oats are grown in all agroclimatic zones. These are one of the most frost-tolerant forage crops. But, it is highly susceptible to climate change during its early growth phases and at anthesis stages.

Alfalfa establishes a symbiotic relationship with a N_2-fixing bacterium (*Sinorhizobium meliloti* L.) providing an extra source of nitrogen (N) for the plant and soil, improving soil structure and increasing soil organic matter (Bourgeois, 1990). Alfalfa is a N_2-fixing plant and it may grow without N fertilizer application. N_2-fixing legume often shows a larger stimulation of growth and photosynthetic rate under elevated CO_2 than species without this capability (Luscher et al., 1998). It has been demonstrated that modulated alfalfa plants are very responsive to elevated CO_2 and their growth being significantly stimulated. According to the results, we can conclude that the dry matter production of the modulated alfalfa plants is limited under drought and elevated temperature conditions, but especially when both stresses are applied simultaneously. Leaf photosynthesis and nodule nitrogen fixation were adversely affected.

22.4 SPECIES MIXTURES

Climate change also exerts a major influence on fodder-forage quality. Grasses contain higher concentrations of fiber and lower concentrations of readily fermentable cell soluble as compared with legumes. Schenk et al. (1997) reported that CO_2 enrichment initially reduced the crude protein content of both species, but eventually increased the crude protein content on a whole sward level as the content of clover increased.

22.5 EFFECTS ON FORAGE QUALITY

Allard et al. (2003) reported that the combined effects of climatic parameters such as elevated CO_2 and temperature on forage quality are likely to be complex and it ranges from positive, neutral, to negative impacts, depending upon, for examples, the accompanying changes in rainfall patterns. Higher ambient temperatures during growth are associated with decreased digestibility, attributable to higher concentrations of cell wall components, and lignin as reported by Hall et al. (2007). A study of ryegrass and orchard grass cultivars indicated that the primary effect of water stress was to increase nutritional value by increasing concentrations of crude protein and digestible neutral detergent fiber. According to the Environmental Protection Agency, an increase in average temperature can lengthen the growing season in regions with relatively cool spring and seasons; adversely affect crops in regions where summer heat already limits production; increase soli evaporation rates, and increase the chances of severe droughts (EPA, 2008a).

22.6 CHANGED CLIMATE: INFLUENCED FORAGE PRODUCTION

1. Climate change-induced shifts in plant species are already under way in rangelands. The establishment of perennial herbaceous species is reducing soil water availability early in the growing season.
2. Backland (2008) reported that higher temperatures will very likely reduce livestock production during the summer season, but these loses will be partially offset by warmer temperatures during winter season.
3. Diseases pressure on crops and domestics animals will likely increase with earlier springs and warmer winters.
4. Projected increase in temperature and lengthening of the growing season will likely extend forage production into late fall and early spring.

22.7 GREENHOUSE GAS EMISSIONS

The primary sources of GHGs in agriculture are the production of nitrogen-based fertilizers, the combustion of fossil fuels such as coal, gasoline, diesel fuel, and natural gas and waste management. Livestock enteric fermentation or the fermentation that takes place in the digestive systems of ruminant animals that results in methane emissions. Carbon dioxide is removed from the atmosphere and converted to organic carbon through the process of photosynthesis. As organic carbon decomposes, it is converted back to carbon dioxide through the process of respiration. Conservation tillage, organic production, cover cropping, and crop rotations can drastically increase the amount of carbon stored in soils.

22.8 CARBON SEQUESTRATION

Carbon sequestration in the agriculture sector refers to the capacity of agricultural lands and forest to remove carbon dioxide from the atmosphere.

Carbon dioxide is absorbed by trees, plants, and crops through photosynthesis and stored as carbon in biomass in forage crops, branches, foliage and roots, and soils (EPA, 2008b). Forests and stable grasslands are referred to as carbon sinks because they can store large amounts of carbon in their vegetation and root systems for long periods of time.

22.9 FORAGE CROPS AND CARBON SEQUESTRATION

High biomass yielding Cumbu Napier hybrid grass is one of the important perennial forage grasses, which is able to increase carbon sequestration through extensive root system and the consequential improvement of soil productivity. An experiment was conducted during 2009 at Department of forage crops, Tamil Nadu Agriculture University (TNAU), to assess the carbon sequestration pattern in Cumbu Napier hybrid grass CO (CN) 4 in clayey loam soil. The green fodder yield, dry matter content, dry fodder yield, and weight of dried leaves were recorded after each harvest. Actually CO_2 absorbed by grasses through photosynthesis is sequestered in the roots and surrounding soil. The compost layer increases the amount of carbon absorbed by the plants and relatives to the amount released back into the atmosphere. The experimental results revealed that the green forage yield GFY of the aboveground portion varied with the climate condition and so with the CO_2 removal (De Luis et al., 2000). The biomass yield of belowground portion and dried leaves was found to be increasing with the age of the crop. It was observed that the total CO_2 removal by the BN hybrid CO (CN) 4 was 153 t/ha/year (aboveground portion 128.7 t/ha/year + belowground portion 1.6 t/ha/year + dried leaves 22.82 t/ha/year).

22.10 METHANE CAPTURE

Large emissions of methane and nitrous oxide are attributable to livestock waste treatment, especially in dairies. Agriculture methane collection and combustion systems include covered lagoons and complete mix and plug flow digesters. Anaerobic digestion converts animal waste to energy by capturing methane and preventing it from being released into the atmosphere. The captured methane can be used to fuel a variety of on-farm applications, as well as to generate electricity.

22.11 MITIGATING THE EFFECT OF CHANGED CLIMATE

Different farming practices and technologies can reduce GHGs emission and prevent changed climate scenario by enhancing carbon storage in soil, preventing existing soil carbon and reducing carbon dioxide, methane, and nitrous oxide emissions.

22.11.1 INNOVATIVE FARMING PRACTICES

Innovative farming practices such as improved cropping systems, conservation tillage, organic farming, land restoration, and nutrient and water management practices are ways that farmers can use to address the climate change. Good management practices have multiple benefits that may also enhance profitability, improve farm energy efficiency, and boost air and soil quality.

22.11.2 CONSERVATION FARMING PRACTICES

These practices generally conserve the soil moisture, improved yield potentiality, and reduce erosion and fuel costs as well as also increase soil carbon status. Examples of practices that reduce carbon dioxide emission and increase soil carbon include direct seeding, field wind breaks, rotational grazing, perennial forage crops, reduced summer fallow, and proper straw management. Using higher yielding crops or varieties and maximizing yield potential can also increase soil carbon.

22.11.3 ORGANIC PRODUCTION SYSTEM

This system increases soil organic matter levels through the use of composted animal manures and cover crops. Organic cropping systems also eliminate the emissions from the production and transportation of synthetic fertilizers.

22.11.4 CONSERVATION TILLAGE

It refers to a number of strategies and techniques for establishing crops in the residues of previous crops, which are purposely left on the soil surface.

Reducing tillage reduces soil disturbance and helps mitigate the release of soil carbon into the atmosphere. Conservation tillage also improves the carbon sequestration capacity of the soil. Conservation tillage also improved water conservation, reduced fuel consumptions, reduced soil erosion, reduced soil compaction, increased planting and harvesting flexibility, reduced labor requirements, and improved soil tilts.

22.11.5 LAND RESTORATION AND LAND-USE CHANGES

Land restoration and land-use changes that encourage the conservation and improvement of soil, water, and air quality as well as typically reduce the GHG emission. Modifications to grazing practices, such as rotational grazing and seasonal use of range land, can lead to GHG reductions. Converting marginal crop land with trees or grass maximizes carbon storage on the land that is less suitable for crops.

22.11.6 GRASSLAND AND CARBON SEQUESTRATION

Grassland is normally considered to sequester carbon. Grassland remaining grassland includes all areas that have been designated as grassland for the past 20 years. Grassland remaining grassland released about 16 MMT CO_2 equiv each year from 2002 to 2008, largely due to droughts causing small losses of carbon per acre over large geographic areas. Land converted to grassland includes all land designated as grassland. Lands are kept in this category for 20 years, after which they are considered grassland remaining grassland. Carbon sequestration was primarily due to conversion of cropland to continuous pasture lands.

22.11.7 GRASS–LEGUME MIXTURE

The growing of leguminous crops as intercrop or rotation crop or as mixed crop in rainfed regions seems to offer best prospect for success as of their complementary functions. Legumes usually maintain their quality better than grasses even at maturity and being rich in protein, which enhance the forage value and also add substantially the much-needed nitrogen to the soil. The grass–legume mixture also improves the physical conditions of the soil, check soil erosion, resist the encroachment of weeds, and withstand the vagaries of weather better than pure stands.

22.11.8 RELEASE OF NEW VARIETIES

Development and release of new varieties for high temperature, drought, and submergence tolerance and evolving varieties which respond positively in growth and yield to high CO_2.

22.11.9 CROPPING SYSTEM-BASED TECHNOLOGY

It will mainly depend on promoting the cultivation of crops and varieties that suited into new cropping systems and seasons, development, and release of varieties with changed duration that can over climate change scenario. Rice–maize–rice bean, rice–oat + lathyrus/pea/gram–maize + cowpea, sorghum + redgram–tomato, sunflower–cowpea–multicut sorghum, baby corn + cowpea–oat–fodder maize + cowpea, maize (as grain or green cob or baby corn)–lucern/berseem–maize + cowpea, etc. food-forage cropping system is found to produce feed (grain/food) and green forage/dry fodder as well as improved soil fertility status and store carbon in the soil. Improved and novel agronomic management practices and crop production technologies, namely, adjustment of sowing dates/planting dates and management of the plant spacing and nutrient management practices may help reduce the adverse effects of changes of climate.

22.11.10 SUSTAINABLE PRODUCTION TECHNOLOGY

Aerobic rice and system of rice intensification is the sustainable rice production technology and rice for the future and promising leguminous forage crops as dual purpose can be grown in rice follows by using residual soil moisture and fertility to improve the soil health and beneficial microbial status and is often grown as intercrop with maize (as baby corn or green cob). Rice bean is also called as "under-utilized" legume or "orphan" crop or poor man's pulse. It can be grown both in uplands and in rice fallows in the summer season. Improved varieties of rice bean, namely, Bidhan rice bean-1, Bidhan rice bean-2, and Bidhan rice bean-3 has been realized for cultivation. Rice bean, *Gliricidia sepium* (gliricidia), *Lablab purpureus* (lablab bean), *Leucaena leucocephala* (subabul), and *Centrosema pubescens* (centro) are examples of drought-tolerant legumes. These type of production technologies and practices have multiple benefits that may also provide food (grain/feed) and forage, enhance profitability, improve soil carbon status as well as soil fertility, and boost air and soil quality.

22.11.11 SURFACE SEEDING OR ZERO TILLAGE

It is one of the important parts of resource conservation technologies. Surface seeding or zero tillage not only restrict the release of soil carbon in the atmosphere but also sometimes helps partially withstand the adverse climate and provides better yield or stabilizes it.

22.12 FUTURE STRATEGIES

1. Development of improved forage varieties of different forage crops with high biomass production, multicut nature, and dual purposes types (seed/grain + green forage/dry fodder).
2. Development of suitable production technologies for forage (fodder)-based cropping system for intensive agriculture and intercropping systems with aims of storing carbon in the soil.
3. Enhancing food cum forage (fodder) production efficiency of grasslands and utilization of wasteland for feed-forage production to mitigate the effects of changed climate.

22.13 CONCLUSION

Agriculture is inherently sensitive to climate variability and its change. The IPCC estimates that the reduction in the options of agricultural GHGs mitigation is cost-competitive with nonagricultural options for achieving long-term climate objectives. From the present study, it may be concluded that there are different ways in agriculture sector to reduce the GHGs emission, namely, land management to increase soil carbon storage, improved food cum forage production system, improved rice cultivation techniques, restoration of degraded lands, improved nitrogen fertilizer application and dedication energy crops, etc. (IPCC, 2007). The issues of climate change and importance of forage crops to mitigate the effects of changed climate scenario have been a major research topic in recent time. We need to emphasis on the potential interactions between the effects of changed climate and ongoing economic interaction.

KEYWORDS

- **forage crops**
- **climate change**
- **production**
- **quality**

REFERENCES

Allard, V.; Newton, P. C. D.; Lieffering, M.; Clark, H.; Matthew, C.; Gray, Y. Nutrient Cycling in Grazed Pastures at Elevated CO_2: N Returns by Animals. *Global Change Biol.* **2003,** *9*, 1731–1742.

Backland, P. Climate Change Science Programme and the Subcommittee on Global Change Research May. *The Effects of Climate Change on Agriculture, Land Resources, Water Resources and Biodiversity in the United States*; Climate Change Science Programme: Washington, DC, 2008.

Barbehenn, R. V.; Chen, Z.; Karowe, D. N.; Spikard, A. C_3 Grasses have Higher Nutritional Quality than C_4 Grasses Under Ambient and Elevated CO_2. *Global Change Biol.* **2004,** *10*, 1565–1575.

Bourgeois, G. Evaluation of an Alfalfa Growth Simulation Model Under Quebec Conditions. *Agr. Syst.* **1990,** *32*, 1–12.

De Luis, I. Efectos del Aumento de la Concentracion de CO_2 Atmosferico en Plantas de Alfalfa Fijadoras de Nitrogeno Bajo Condiciones de Estres. *Tesis Doctoral*, Universidad de Navarra, Pamplona, Spain, 2000.

Drake, B. G.; Gonzalez-Meler, M. A.; Long, S. P. More Efficient Plants: A Consequence of Rising Atmospheric CO_2. *Plant Mol. Biol.* **1997,** *48*, 609–639.

Hall, A. E.; Allen, L. H. Jr. Designing Cultivars for the Climatic Conditions of the Next Century. *International Crop Science Society*; Guilford Road, Madison, WI-53711, 2007.

IPCC. *Climate Change 2007. The Physical Science Basis.* Contribution of Working Group I to the Fourth Assessment Report of the Intergovernmental Panel on Climate Change, Cambridge University Press: Cambridge, 2007.

Kimball, B. A. Carbon Dioxide and Agricultural Yield: An Assemblage and Analysis of 430 Prior Observations. *Agron. J.* **1983,** *75* (5), 779–788.

Long, S. P.; Ainsworth, E. A.; Leakey, A. D. B.; Nosberger, J.; Ort, D. R. Food for Thought: Lower-than-expected Crop Yield Stimulation with Rising CO_2 Concentrations. *Science* **2006,** *312* (5782), 1918–1921.

Luscher, A.; Hendrey, G. R.; Nosberger, J. Long-term Responsiveness to Free Air CO_2 Enrichment of Functional Types, Species and Genotypes of Permanent Grasslands. *Oecologia* **1998,** *113*, 37–45.

Saeed, I. A. M.; Nadia, A. H. El. Forage Sorghum Yield and Water Use Efficiency Under Variable Irrigation. *Irrig. Sci.* **1998,** *18*, 67–71.

Schenk, U.; Jager, H. J.; Weigel, H. J. The Response of Perennial Rye Grass/White Clover Swards to Elevated Atmospheric CO_2 Concentrations. *New Physiol.* **1997,** *135*, 67–69.

Tans, P.; Keelinng, R. Scripps Institution of Oceanography, Measurement at Mauna Loa, Hawaii by the NOAA (accessed Oct 14, 2012).

Wheeler, T. R.; Craufurd, P. Q.; Ellis, R. H.; Porter, J. R.; Vara Prasad, P. V. Temperature Variability and the Yield of Annual Crops. *Agric. Ecosyst. Environ.* **2000,** *82*, 159–167.

INDEX

H

J

T